Manar Kassou

The impact of Blockchain on Big Data

Manar Kassou

The impact of Blockchain on Big Data

A Technological Revolution in Data Management and Analysis

ScienciaScripts

Imprint

Cover image: www.ingimage.com

This book is a translation from the original published under ISBN 978-620-6-71476-7.

Publisher:
Sciencia Scripts
is a trademark of
Dodo Books Indian Ocean Ltd. and OmniScriptum S.R.L publishing group

120 High Road, East Finchley, London, N2 9ED, United Kingdom
Str. Armeneasca 28/1, office 1, Chisinau MD-2012, Republic of Moldova, Europe
Printed at: see last page
ISBN: 978-620-8-01999-0

THE IMPACT OF BLOCKCHAIN ON BIG DATA

KASSOU Manar

Dedicated to

"For all the altruistic souls around the world who offer their help, guidance and motivation so that others can succeed in life. For the exceptional people who believe in humanity and kindness, and for all those who love others regardless of color or religion. "

Table of contents

INTRODUCTION

Today, Blockchain is considered the fifth evolution of information and communication technologies (ICT) by many experts. This assertion is supported by various elements such as the rapid evolution of digital infrastructures, the exponential increase in data volume and the growing need for security and transparency in digital transactions. A study conducted by Price Waterhouse Coopers (2018) among 600 business leaders in 15 different countries revealed that 84% of the companies surveyed planned to use Blockchain at some point in their operations.

These companies perceive Blockchain as a disruptive technology capable of having a significant impact on their business sectors, notably by reducing costs, improving process efficiency, and increasing the transparency and security of transactions.

Blockchain has gained considerable recognition from researchers and is attracting increasing attention in the field of global innovation. Described as a "trust machine", it was predicted to "redefine the world" by The Economist in 2015.

This technology is often compared to other next-generation technologies such as the Internet of Things (IoT), cloud computing and Big Data.

Blockchain is distinguished by its unique and innovative application model, which combines distributed data storage, decentralized and independent peer-to-peer transactions, automatic and intelligent consensus mechanisms, information management systems, programmable smart contracts and dynamic encryption algorithms (Kassou et al., 2020).

According to the Gartner report (2016-2017), Blockchain has been ranked among the emerging technologies most prone to inflated expectations. It enables bilateral and multi-party transactions in a distributed and decentralized environment, while offering features such as complete network recording, provenance of information and resistance to forgery (Kassou et al., 2020). In addition, the Blockchain infrastructure enables rules to be enforced in the event of a breach of trust, which is crucial in a society where a large proportion of interactions rely on trust and rule enforcement.

The social and economic implications of Blockchain applications are vast and potentially polarizing, as this technology could profoundly transform the way we structure value-based transactions and society itself (Al-Saqaf & Seidler, 2017).

As Blockchain continues to develop and deploy on a large scale, it promises to become a major driving force in the evolution of digital systems and business practices around the world.

One of the most intriguing aspects of Blockchain is its ability to provide innovative solutions to traditional data management challenges. For example, in the healthcare sector, Blockchain can ensure the confidentiality and security of patient medical records while enabling the rapid and secure sharing of information between healthcare professionals. In finance, it can facilitate faster and cheaper cross-border transactions, while reducing the risk of fraud thanks to its decentralized verification mechanisms.

Blockchain also has transformative potential in supply chain management. By offering complete traceability of products from origin to final destination, this technology can improve transparency, reduce fraud and increase the efficiency of logistics processes.

From food companies to luxury goods manufacturers, many are already exploring the benefits of Blockchain to guarantee the authenticity and provenance of their products. In governance and public services, Blockchain could revolutionize identity management, the conduct of elections and the delivery of administrative services. By securing data and making processes more transparent, this technology could boost citizens' trust in public institutions and reduce corruption.

However, the adoption of Blockchain is not without its challenges. Businesses and governments must overcome technical, regulatory and cultural hurdles to fully integrate this technology. The complexity of implementing Blockchain-based systems requires specialized skills and a robust infrastructure. In addition, the need for appropriate standards and regulatory frameworks is essential to ensure the smooth and secure adoption of Blockchain on a large scale. In conclusion, Blockchain represents a major evolution in today's technological landscape. Its ability to offer secure, transparent and efficient solutions for managing data and transactions makes it a key technology for the future. By exploring and fully exploiting the potential of Blockchain, companies and organizations can not only improve their efficiency and competitiveness, but also contribute to the creation of a more equitable and reliable digital environment. This integration paves the way for a new era of innovation where the management of massive data is both reliable and secure, opening up unexplored horizons for research and industry.

CHAPTER 1

"The emergence of Blockchain"

Advances in computer science have fostered the development of modern networks and cryptography, both of which are essential to the emergence of blockchain technology. Cryptography plays a crucial role in blockchain. However, it's important to point out that the origins of cryptography go back long before the introduction of computing. Indeed, the first example of an encrypted document to be discovered was a clay tablet unearthed in Iraq, dating back to the 16th century.

A potter had written a recipe on it, deleting certain consonants and changing the spelling of words. Various encryption techniques were developed over time, culminating in the famous Enigma machine[15] used by the Germans during the Second World War.

After the Second World War, Γ era of modern communication began, introduced by Claude Shannon and his paper published in 1948: "A Mathematical Theory of Communication". By theorizing the digitization of communications, Claude Shannon paved the way for modern cryptology. Some twenty years later, in 1969 in the United States, the ARPANET project initiated by the U.S. Defense Department and universities (UCLA, Stanford) led to the creation of the first computer-to-computer communications network.

It was from this network that the Internet was born, a tool for hosting multiple applications such as e-mail, the web and peer-to-peer networks, the latter being the basis for the construction of a network interconnecting Blockchain users. In the mid-1970s, asymmetric cryptography began to expand. The algorithms derived from it enable two entities to exchange encryption keys to solve two crucial problems:

- ✓ Ensure the confidentiality of a communication between two parties,
- ✓ Guarantee the authenticity of the sender of a message.

(Leslie Lomport, 1982), an American computer scientist, published "The Byzantine General Problem". The Byzantine dilemma is a mathematical metaphor for assessing a system's resistance to communication failures and the integrity of its participants. This assessment is of paramount importance in a network open to the public, and susceptible to malicious or fraudulent users.

David Chaum (1982) created Ecash, a centralized electronic currency marketed by Digicash in 1989. This currency guarantees anonymous payments based on a cryptographic protocol invented by David Chaum himself (the blind signature). Although Digicash went bankrupt in 1998, the cryptographic innovation of the blind signature continues to be used, notably in electronic voting systems.

The introduction of the Ecash digital currency was a crucial step towards the development of virtual currencies, such as Bitcoin, which became the first large-scale use of blockchain technology. The first notion similar to Blockchain and the mention of "distributed trust" appeared in 1991.

The article "How to Time-Stamp a Digital Document" by (Haber and Stornetta), published in the Journal of Cryptography, examines how it is possible to time-stamp any digital content without the possibility of fraud as to the file's content. In the 90s, in the era of the rise of the digital format, these two researchers wondered how to provide a dated proof of existence for a digital document, with the aim of guaranteeing the primacy of a scientific discovery.

In addition to attesting the date of deposit of a document, their research objective was to set up a system for asserting the unalterable integrity of the deposited document. Thus, they proposed to use computer hash functions to generate a tamper-proof fingerprint of a file. For the first time, they proposed to link data chronologically to create a system in which it would be impossible to alter the sequence of events.

This is how the concept of blockchain was first introduced. The notions of hashing and blockchain, which are essential elements of blockchain technology, were largely taken up almost 20 years later by Bitcoin.

Adam Back (2002), a British cryptographer, proposed the HashCash proof-of-work system. This system was developed in response to the problem of spam and denial-of-service attacks.

Adam Back's proof-of-work principle has become an essential element of the Bitcoin blockchain, enabling new data to be validated. (Nick Szabo, 1998-2005) developed the BitGold project.

The technical architecture underlying the operation of this currency is very similar to that of Bitcoin. For the first time, a decentralized digital currency was brought to market. Like Bitcoin after it, BitGold uses a proof-of-work system, with transactions organized in blocks and cryptographically linked to each other, then distributed across a network.

Although BitGold is very similar to Bitcoin, there is still a technological barrier. Indeed, this currency offers a highly vulnerable solution to the problem of double spending. All these technological innovations and projects, more or less successful, have helped to create a fertile environment for the Blockchain. Indeed, we shall see that the first Bitcoin Blockchain can be technologically deconstructed as the sum of the above innovations.

1.1. THE RISE OF BITCOIN

The advent of Bitcoin marks a major revolution in financial technology and digital payment systems. In 2009, the mysterious Satoshi Nakamoto introduced Bitcoin, a decentralized digital currency based on an innovative technology called blockchain.

This technology is based on advanced cryptographic principles and a peer-to-peer network architecture, enabling secure, transparent transactions with no trusted intermediaries.

The rise of Bitcoin did not happen in isolation; it is the result of decades of research and innovation in cryptography, computer science and communication networks. Bitcoin emerged against a backdrop of shaken confidence in traditional financial institutions, particularly in the wake of the 2008 global financial crisis.

The Cypherpunks, a group of cryptographers and activists, have played a crucial role in promoting privacy and decentralization through cryptographic technologies. Their vision of a world where trust is based on code and mathematics rather than centralized institutions has found concrete application in Bitcoin.

This introduction of the first cryptocurrency not only paved the way for many other digital currencies, but also initiated a profound transformation of economic systems and transaction mechanisms. Bitcoin's impact extends beyond the simple transfer of value; it proposes a new form of economic interaction based on decentralization, transparency and security.

Studying the advent of Bitcoin provides an insight into the technological and philosophical underpinnings of this innovation. By exploring the contributions of the Cypherpunks, advances in cryptography and the evolution of computer networks, this section highlights how Bitcoin was able to emerge as a response to the need for a reliable digital currency independent of traditional institutions.

1.1.1. CYPHERPUNKS

The Cypherpunks movement takes its name from the combination of the words "Cyberpunk", evoking the digital universe, and "cipher" meaning "cipher" or "secret code". This group, made up mainly of cryptographers, mathematicians and computer scientists, emerged following the publication of David Chaum's articles in the late 1980s. The Cypherpunks see the digital revolution as a dual reality: an opportunity and a threat.

They believe that the digitization of communications enables governments to conduct large-scale surveillance of information exchanges. In 1992, Eric Hughes, Timothy C. May and John Gilmore met regularly and created a mailing list that by 1997 had grown to 2,000 subscribers. This mailing list enabled them to communicate and collaborate on common projects. In 1993, Eric Hughes drew up the "Cypherpunk Manifesto", setting out the movement's key demands and objectives. The three main axes developed are:

Figure 1: The manifesto's main themes

Source : created by the author

At the heart of Cypherpunk's philosophy is the conviction that the big question in Γ Internet age is whether Γ State will strangle individual freedom and privacy through its capacity for surveillance. These activists act on two fundamental principles.

The first is to contribute to technological advances, particularly in cryptography. The second is to integrate cryptography into various online applications. Through their reflections, the Cypherpunks laid the foundations for a movement which, a few years later, would be at the origin of the creation of Bitcoin.

As early as 1998, Wei Dai published on the Cypherpunks mailing list the foundations of a currency called B-money, which was intended to be anonymous, electronic and decentralized. In their view, the point of having a decentralized currency was to free oneself from the control exercised by central institutions, such as banks.

In pursuing their goals, this movement developed a concept that is essential to blockchain. According to the Cypherpunks, since a central state cannot be trusted absolutely to guarantee, among other things, the privacy of citizens, trust must be based on the software itself and the computer code it contains, rather than on legal regulations. In other words, this vision marks a transfer of trust from the state and central authorities to the infrastructures used for exchanges.

1.1.2. Blockchain 1.0: Bitcoin

The chronology of blockchain can be segmented into two main periods, divided by the creation of Bitcoin in 2009, which is closely linked to its development.

The period from the 1970s to 2009 represents not only the emergence of the various technologies that would later form the Bitcoin Blockchain, but also the intention of certain players to combine them to create systems that increasingly resemble today's Blockchain model.

Bitcoin was the first project to put this technology into practice on a large scale. On January 3, 2009, the first block of the Bitcoin blockchain was created by Satoshi Nakamoto, giving birth to the first digital currency based on blockchain technology.

The validation of the Bitcoin concept then inspired a host of programmers and entrepreneurs to extend the use of Blockchain beyond cryptocurrencies. Today, the first projects in the healthcare field are beginning to be developed.

The mysterious Satoshi Nakamoto reserved the bitcoin.org domain name on August 19, 2008, but very little is known about this person. No one can say for sure whether Satoshi Nakamoto is a single person or a group of people working together.

On October 31 of the same year, Satoshi Nakamoto posted a message on a cryptographic mailing list, similar to the one used by Cypherpunks: "I've been working on a new electronic payment system that is entirely peer-to-peer, with no trusted third parties."

Main properties :

Figure 2: Main properties of Bitcoin

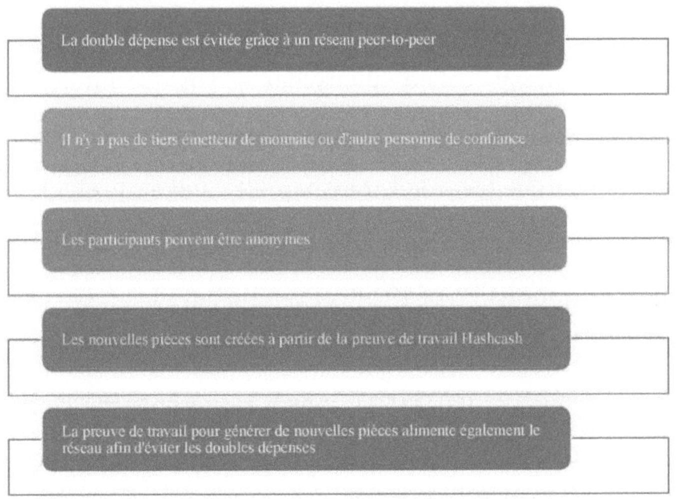

Source : created by the author

The Bitcoin White Paper (Nakamoto, 2008) details the objectives and technical means of creating this electronic currency. This document marks a crucial step in the emergence of a fully decentralized currency, with no trusted intermediaries and an adequate level of security.

Bitcoin is widely regarded as the first cryptocurrency, and paved the way for many others to follow. Each cryptocurrency has its own specific technical characteristics and objectives.

As of August 19, 2018, there were more than 1,600 different cryptocurrencies. Despite their diversity, all these cryptocurrencies share a common characteristic: they are based on a technology called Blockchain. This technology has revolutionized the way crypto-currencies work, enabling transactions to be recorded transparently and securely. January 3, 2009 marked the creation of the first block of bitcoins, followed by the first transaction of 10 bitcoins nine days later. On October 5, 2009, the first estimate of bitcoin's value was made, based on its production cost (€0.00071), which represents the electricity required to maintain the network. In November 2010, the value of bitcoin reached 40 cents, while on December 12, 2010, Satoshi Nakamoto announced his departure from the Bitcoin project.

At the end of 2013, the value of bitcoin exceeded 800 euros, marking its entry into the mainstream and acceptance by the first organizations. Authorities such as governments and central banks are also beginning to take an interest in this cryptocurrency.

In 2017, bitcoin reached its all-time peak with a value of 16,323 euros and a market capitalization of over 281 billion euros. Since then, its value has declined, standing at around 5,600 euros in September 2018. Satoshi Nakamoto was able to capitalize on technological advances such as cryptography and proof-of-work, while building on previous projects such as E-cash and B-money.

Bitcoin and its underlying technology, Blockchain, have revolutionized trust between parties who don't know each other, enabling exchanges of value without the intervention of a trusted third party. Bitcoin was the first use case for Blockchain technology, paving the way for its diversification and growing maturity. The second period, post-2009, is marked by an explosion of innovations and applications based on blockchain. Entrepreneurs and developers quickly recognized blockchain's potential beyond cryptocurrencies, exploring areas such as smart contracts, decentralized finance (DeFi), supply chain management, and even creative industries like art and music.

Ethereum, launched in 2015 by Vitalik Buterin, is an emblematic example of this diversification. Ethereum introduced the possibility of creating and executing smart contracts, autonomous programs that run automatically when predefined conditions are met. This innovation has paved the way for a host of new applications, enabling the design of complex decentralized systems that are not dependent on any central authority. Blockchain has also begun to attract the attention of traditional companies and governments. Technology giants such as IBM and Microsoft have developed blockchain platforms for businesses, enabling transparent and secure supply chain solutions.

Governments, meanwhile, are exploring the use of blockchain for applications such as secure electronic voting, land registration and digital identity management. In parallel, decentralized finance (DeFi) has emerged as a particularly dynamic field. DeFi applications seek to recreate and enhance traditional financial services - lending, borrowing, trading, insurance - using decentralized protocols on the blockchain. This promises greater accessibility, transparency and security, while reducing the costs and inefficiencies associated with traditional financial intermediaries. Finally, the social and cultural impact of blockchain should not be underestimated. The concept of digital property, notably through non-fungible

tokens (NFTs), has begun to transform sectors such as art, music and video games, offering new ways of monetizing and protecting intellectual property. NFTs enable creators to sell digital artworks with proof of ownership and authenticity registered on the blockchain, creating new opportunities for artists and collectors. The post-2009 period has seen blockchain evolve from a niche technology focused on cryptocurrencies to a versatile and disruptive infrastructure capable of transforming many aspects of our society and economy.
This rapid and ongoing evolution underlines the immense potential of blockchain and its long-term implications for a variety of industries and social practices.

1.2. After Bitcoin

The creation of Bitcoin in January 2009 sparked growing interest in the underlying technology, the Blockchain. Over the years, this technology has attracted developers who have used it to create new cryptocurrencies. At the same time, developers have also been looking at the many applications of Blockchain technology outside the realm of digital currencies. In an article for the Financial Times, Sally Davies compared the Blockchain to the internet, highlighting its potential as an electronic system enabling the construction of a variety of applications, with currency being just one facet. As bitcoin's Blockchain is open source, it has been adapted by various people to create numerous cryptocurrencies, although only a few of the more than 1,600 in existence will be mentioned. These alternative cryptocurrencies, known as altcoins, generally bring incremental improvements rather than disruptive innovations. Litecoin, for example, which appeared in October 2011, has reduced transaction confirmation times compared to bitcoin. Peercoin, meanwhile, introduced a new consensus mode called "proof of stake" in August 2012, which consumes less energy than the proof of work used by bitcoin. Monero and ZCash, created in 2014 and 2016 respectively, have improved transaction confidentiality by limiting access to the transaction list. Finally, BitcoinCash, derived from bitcoin in August 2017, increased the size of transaction blocks to solve scalability issues. Thus, blockchain, which was born with bitcoin, has paved the way for a multitude of cryptocurrencies, each with specific features and functionalities, testifying to the diversity and ongoing evolution of this revolutionary technology.

CONCLUSION

This first chapter traced the emergence of blockchain technology and the advent of Bitcoin, highlighting the historical, technological and philosophical elements that led to this digital revolution. We saw how advances in cryptography, innovations in computer networking, and the contributions of the Cypherpunks laid the groundwork for the creation of Bitcoin. The history of cryptography, from the earliest encryption techniques to modern algorithms, has shown the importance of protecting information in an increasingly digital world. The key concepts introduced by the pioneers of asymmetric cryptography and their solutions to communication and security problems have directly influenced the development of blockchain. The work of Claude Shannon, Leslie Lamport, David Chaum and many others has led to the conceptualization and implementation of secure, decentralized systems. The Cypherpunks initiative, with its vision of a digital society where privacy and decentralization are paramount, has been crucial in creating a favorable environment for the development of cryptocurrencies. Their discussions, manifestos and projects helped to devise alternatives to traditional financial systems, culminating in the creation of Bitcoin. With the introduction of Bitcoin in 2009, Satoshi Nakamoto not only offered a new form of digital currency, but also proposed an underlying technology - the blockchain - that has the potential to transform many sectors beyond financial transactions alone. Bitcoin, with its properties of decentralization, security and transparency, has demonstrated that trust can be transferred from centralized institutions to robust cryptographic protocols. In conclusion, the advent of Bitcoin and the emergence of blockchain represent a significant milestone in the evolution of digital technologies. They mark the beginning of a new era in which decentralization and cryptography play a central role in the way we conceive economic transactions and interactions. Bitcoin has paved the way for a multitude of innovations and potential applications of blockchain, whose impacts will be felt in various areas of society.

CHAPTER 2

"Smart contract, Ethereum and Hyperledger Fabric/'

Since the introduction of Bitcoin in 2009, blockchain technology has evolved rapidly, profoundly transforming transaction systems and economic interactions. This chapter focuses on one of the most significant advances in blockchain: smart contracts. By harnessing the power of blockchain to automate and secure agreements, smart contracts are opening up new perspectives for many industries.

1.1. Blockchain 2.0: Smart Contracts

Smart contracts represent a major step forward in the evolution of blockchain, often referred to as Blockchain 2.0. Unlike simple bitcoin transactions, smart contracts make it possible to integrate detailed and complex instructions directly into the blockchain.

In essence, a smart contract is software code that executes autonomously and automatically, without human intervention, thus guaranteeing faithful execution of the terms agreed between the parties.

This capacity for self-application and decentralization eliminates the need for trust between the agents involved, and enhances the transparency and security of transactions. A classic illustration of this technology is that of vending machines, which operate algorithmically to execute transactions reliably and repeatedly.

The design of smart contracts is based on fundamental principles of cryptography and distributed algorithms. They operate on a decentralized network of nodes, where each node executes the smart contract code independently, guaranteeing consistent and transparent execution. This distributed nature eliminates single points of failure and ensures greater system resilience. Smart contracts offer several notable advantages:

1. **Automation and efficiency**: Smart contracts can automate complex processes, reducing operational costs and minimizing human error. For example, in the insurance sector, smart contracts can automatically process claims and make payments according to predefined conditions.

2. **Transparency and immutability**: All smart contract transactions and conditions are recorded transparently on the blockchain. Once recorded, this information cannot be altered, guaranteeing data integrity and traceability.

3. **Security**: Thanks to the use of cryptography, smart contracts offer a high level of security, making attempts at

manipulation or fraud extremely difficult. The decentralized nature of the network further enhances this security by eliminating single points of failure.

4. **Reliability and guaranteed execution**: smart contracts execute automatically as soon as predefined conditions are met, eliminating the delays and uncertainties associated with manual processes.

A concrete example of the application of smart contracts is in the field of financial services. Decentralized finance platforms (DeFi) use smart contracts to provide services such as lending, borrowing, and trading without intermediaries. These contracts enable users to interact directly with each other, creating a more open and inclusive financial ecosystem.

Smart contracts are also finding applications in other sectors such as supply chain management, where they can automatically track and verify the provenance of products, and real estate, where they can facilitate property transactions by automating title transfers.

However, despite their many advantages, smart contracts also present challenges and limitations. One of the main challenges is the complexity of the code and the risk of bugs or vulnerabilities.

Errors in smart contract code can have serious consequences, as demonstrated by the DAO (Decentralized Autonomous Organization) incident in 2016, where a flaw in a smart contract led to the loss of millions of dollars in Ether.

Moreover, the legal and regulatory aspects of smart contracts remain unclear. The legal recognition of these contracts and their compliance with existing laws vary from one jurisdiction to another, which may pose obstacles to their widespread adoption.

In conclusion, smart contracts represent a transformative innovation in blockchain, offering reliable and transparent transaction automation. Their application potential extends to many sectors, promising improved process efficiency and security. However, to fully realize this potential, it is necessary to overcome the technical and legal challenges that accompany this emerging technology.

1.2. Ethereum

Ethereum, launched in 2015 by Vitalik Buterin, represents a major breakthrough in blockchain technologies. Unlike Bitcoin, which focuses primarily on financial transactions, Ethereum is designed as a global platform for the development and execution of smart contracts and decentralized applications (dApps). This versatility is made possible by the Ethereum Virtual Machine

(EVM), a runtime environment that enables anyone to write and deploy executable code on the blockchain.

Ethereum's blockchain programming language, called Solidity, is Turing-complete, meaning it can perform any computational operation that could be performed with a traditional computer. This enables developers to create highly sophisticated smart contracts and build complex applications that run autonomously on the blockchain network. This capability has given rise to a host of innovations and positioned Ethereum as the benchmark platform for blockchain projects.

One of the most revolutionary aspects of Ethereum is its ability to support decentralized autonomous organizations (DAOs). DAOs are organizations that operate without centralized management, thanks to smart contracts that dictate operating rules and make decisions based on the consensus of participants. This structure makes it possible to create truly decentralized organizations that are transparent and resistant to censorship.

Ethereum 2.0

Due to the limitations of the initial version of Ethereum, particularly in terms of scalability and energy efficiency, the Ethereum community has launched an ambitious upgrade project known as Ethereum 2.0. This upgrade aims to transform

Ethereum from a Proof of Work (PoW)-based consensus mechanism to a Proof of Stake (PoS) mechanism.

The transition to Ethereum 2.0 takes place in several phases:

1. **Phase 0: Beacon Chain -** Launched in December 2020, the Beacon Chain is a new blockchain that introduces the PoS consensus mechanism. It runs parallel to the existing Ethereum blockchain and lays the foundations for future upgrades.
2. **Phase 1: Shard Chains -** This phase will introduce the concept of sharding, which divides the blockchain into several segments (shards) to increase transaction processing capacity and improve network scalability. Each shard operates as an independent blockchain, but is secured by the Beacon Chain.
3. **Phase 1.5: Docking -** In this phase, the current Ethereum blockchain will merge with the Beacon Chain, completely transforming Ethereum into a PoS network and eliminating the need for PoW.
4. **Phase 2: eWASM -** The final phase aims to further enhance Ethereum's performance by introducing a new virtual machine, Ethereum WebAssembly (eWASM), which will replace the EVM and enable even faster and more efficient smart contract executions.

➢ Ethereum use cases

One of Ethereum's most popular applications is in the field of decentralized finance (DeFi). DeFi platforms use smart contracts to create open financial services that operate without intermediaries. For example, users can lend and borrow digital assets, trade cryptocurrencies, and earn interest on their funds via automated protocols. One of DeFi's most notable projects is MakerDAO, which enables the creation of stablecoin DAI, a cryptocurrency whose value is indexed to the US dollar and which is generated and regulated by smart contracts.

Ethereum has also paved the way for the rise of non-fungible tokens (NFTs), unique digital assets that can represent works of art, collectibles, or elements of video games. NFTs have become extremely popular, enabling artists and creators to sell and protect their work in a decentralized way. Ethereum's ERC-721 standard defines the rules for NFTs, guaranteeing their compatibility with various markets and applications.

Another area of innovation on Ethereum is digital identity and personal data management. Projects like uPort enable users to manage their digital identities securely and privately, controlling access to their data without having to depend on centralized service providers.

- **Challenges and prospects**

Despite its many advantages, Ethereum faces a number of challenges. Scalability remains a major issue, with the network often congested by high transaction volumes, resulting in high transaction fees. The transition to Ethereum 2.0 and the introduction of sharding are expected to alleviate these problems, but the process is complex and will take time to fully realize. In addition, the security of smart contracts is an area of constant concern. Coding errors or vulnerabilities can lead to significant financial losses, as was the case with the DAO incident. It is therefore crucial to develop best security practices and formal verification tools for smart contracts.

In conclusion, Ethereum has transformed the blockchain landscape by offering a flexible and powerful platform for the development of smart contracts and decentralized applications. Its continued evolution with Ethereum 2.0 and growing adoption in various sectors testify to its long-term potential to revolutionize many aspects of our economy and society.

1.3. Hyperledger Fabric

Hyperledger Fabric, developed by the Linux Foundation in collaboration with IBM, represents a major breakthrough in the field of enterprise blockchains. This private, open-source platform is designed to meet the specific needs of enterprises wishing to take advantage of the benefits of blockchain while retaining full control over transaction confidentiality and visibility. Unlike public blockchains such as Bitcoin or Ethereum, Hyperledger Fabric enables tailor-made solutions to be created to suit the requirements of various industry sectors.

One of the main advantages of Hyperledger Fabric is its modularity. The platform offers a flexible architecture that allows developers to choose and configure various components according to their specific needs.

This includes modules for identity management, access policies, consensus mechanisms, and blockchain services. This flexibility enables companies to create blockchain networks that meet their unique performance, security and compliance requirements.

1.3.1. CONSENSUS MECHANISM

Hyperledger Fabric uses a sophisticated consensus mechanism called Practical Byzantine Fault Tolerance (PBFT). This mechanism is designed for distributed networks, and enables operations to be maintained even in the presence of a certain fault rate. PBFT ensures that all network participants reach a consensus on the state of the ledger, thus guaranteeing the reliability and integrity of transactions.

Unlike the proof-of-work (PoW) mechanisms used by Bitcoin, PBFT is more efficient in terms of power consumption and transaction processing speed, making it particularly suitable for enterprise environments.

1.3.2. TRANSACTION CONFIDENTIALITY

One of the strengths of Hyperledger Fabric lies in its ability to manage transaction confidentiality. Companies can configure their transactions to be public or confidential, depending on the nature of the information exchanged.

This feature is particularly important for sectors that handle sensitive data, such as finance, healthcare or logistics. Hyperledger Fabric enables private channels to be defined, where only authorized parties can access specific information, ensuring protection of intellectual property and compliance with data privacy regulations.

1.3.3. SMART CONTRACTS AND CHAINCODES

Hyperledger Fabric supports smart contracts, called chaincodes in this platform. Chaincodes are programs running on the blockchain that automate and secure business processes. They enable complex business rules to be defined and executed seamlessly and efficiently.

Chaincodes can be written in common programming languages such as Go or Java, making them easy for developers to adopt. Thanks to chaincodes, companies can automate processes such as order management, payment verification and shipment tracking, reducing operational costs and minimizing the risk of human error.

1.3.4. Applications and Use Cases

Hyperledger Fabric is used in a variety of sectors to create blockchain solutions tailored to specific business needs. Here are some examples of use cases:

1. **Supply chain management**: Hyperledger Fabric enables the creation of transparent and secure supply chain networks. Companies can track products at every stage of the chain, from production to delivery, ensuring the authenticity and traceability of goods. This is particularly useful for industries such as agri-food, where product provenance and quality are crucial.
2. **Finance and banking**: Financial institutions use Hyperledger

Fabric to automate settlement and clearing processes, improve transparency and reduce costs. The platform also enables the creation of secure and efficient interbank payment networks, facilitating cross-border transactions.

3. **Healthcare**: In the healthcare sector, Hyperledger Fabric is used to manage electronic medical records securely and confidentially. Patients can authorize access to their medical records to specific healthcare professionals, ensuring the confidentiality and integrity of medical information.
4. **Identity management**: Hyperledger Fabric enables the creation of secure digital identity management systems. Companies can verify user identity and control access to resources in a decentralized way, reducing the risk of fraud and identity theft.

1.3.5. CHALLENGES AND PROSPECTS

Despite its many advantages, Hyperledger Fabric faces a number of challenges in achieving wider adoption. One of the main challenges is the complexity of implementation. The platform's flexibility and modularity require in-depth technical expertise to effectively configure and manage blockchain networks.

In addition, managing transaction confidentiality and complying with local regulations can pose additional challenges for companies operating in multiple jurisdictions. In addition, competition with other blockchain platforms, such as Ethereum and Corda, poses challenges in terms of market share and ongoing innovation.

Hyperledger Fabric must continue to evolve and adapt to changing business needs to maintain its leadership position in the enterprise blockchain field. In conclusion, Hyperledger Fabric offers a powerful and flexible solution for companies wishing to exploit the benefits of blockchain technology.

Its ability to manage confidential transactions, automate business processes via chaincodes, and provide a secure and transparent environment makes it an ideal platform for a wide range of industrial applications. As the technology continues to develop, Hyperledger Fabric is well positioned to play a central role in the digital transformation of enterprises.

1.4. BLOCKCHAIN 3.0

Blockchain 3.0 marks a new stage in the evolution of blockchain technologies, further expanding the possibilities and applications of this technology beyond cryptocurrencies and smart contracts. This new generation of blockchain aims to resolve the limitations of previous versions in terms of scalability, interoperability, sustainability and governance, while exploring new fields of application that touch almost every aspect of human and industrial activity.

1.4.1. SCALABILITY AND PERFORMANCE

One of the main criticisms levelled at first- and second-generation blockchains is their lack of scalability. Blockchains such as Bitcoin and Ethereum have struggled to handle high transaction volumes, resulting in high fees and slow confirmation times. Blockchain 3.0 introduces innovative solutions to improve scalability and network performance. These include sharding, sidechains, and Layer 2 technologies such as state channels and Lightning networks.

- **Sharding**: This technique divides the blockchain into smaller segments called shards, each capable of processing transactions in parallel. This considerably increases transaction processing capacity without compromising security or decentralization.

- **Sidechains**: Sidechains are parallel blockchains that interact with the main blockchain. They offload some of the work from the main blockchain, improving the efficiency and speed of transactions.

- **Layer 2 technologies**: Layer 2 solutions, such as state channels and Lightning networks, enable off-chain transactions to be processed, reducing congestion on the main blockchain while maintaining transaction security and integrity.

1.4.2. INTEROPERABILITY

Blockchain 3.0 also aims to improve interoperability between different blockchains. In previous generations, each blockchain often operated as a silo, unable to communicate or interact with other blockchains. The new generation of blockchain introduces interoperability protocols that enable blockchains to share data and assets seamlessly and securely.

- **Polkadot protocol**: Polkadot is an interoperability protocol that connects several specialized blockchains into a single, unified metachain. This structure enables blockchains to transfer messages and assets securely between each other.

- **Cosmos**: Cosmos is another project aimed at creating a blockchain interoperability network, enabling different blockchains to interconnect via the Inter-Blockchain Communication (IBC) protocol. Cosmos aims to create an Internet of blockchains, where blockchains can interact in a fluid, decentralized way.

1.4.3. Sustainability and Eco-Efficiency

With growing awareness of environmental impacts, Blockchain 3.0 is also focusing on sustainability and energy efficiency. Early generations of blockchain, particularly those using the proof-of-work (PoW) mechanism, were criticized for their excessive energy consumption. Blockchain 3.0 explores greener consensus mechanisms such as proof-of-stake (PoS) and proof-of-authority (PoA).

- **Proof of Stake (PoS)**: This consensus mechanism replaces miners with validators who are chosen according to the amount of crypto-currency they hold and are willing to "stake" to validate transactions. This considerably reduces energy consumption compared with proof-of-work.
- **Proof of Authority (PoA)**: In this model, validators are pre-approved entities that are responsible for validating transactions. Proof of authority is less energy-intensive and offers high efficiency for private blockchain networks or consortia.

1.4.4. Decentralized Governance

Governance is a crucial aspect of Blockchain 3.0, aiming to offer more democratic and decentralized decision-making structures. Decentralized governance mechanisms enable blockchain participants to vote on protocol updates, rule changes, and other important decisions.

- **DAO (Decentralized Autonomous Organizations)**: DAOs are organizations run by smart contracts where decisions are made by members in a decentralized way. DAOs can manage funds, vote on proposals, and operate without direct human intervention, ensuring greater transparency and accountability.
- **Tezos**: Tezos is a self-amending blockchain that allows token holders to propose and vote on changes to the protocol, thus offering continuous and scalable governance of the network.

1.4.5. EMERGING APPLICATIONS AND USE CASES

Blockchain 3.0 explores a wide range of new applications beyond the traditional financial and technological domains. These applications cover sectors such as healthcare, energy, natural resource management, and many others.

- **Healthcare**: blockchains can be used to manage electronic medical records, ensure drug traceability, and facilitate clinical trials securely and transparently.

- **Energy**: Blockchain solutions can be used to manage decentralized energy distribution networks, facilitate peer-to-peer energy trading, and track renewable energy certificates.
- **Natural resource management**: blockchains can help track and verify the origin and supply chain of natural resources, such as minerals and agricultural products, ensuring sustainable and ethical management.

1.4.6. CHALLENGES AND PROSPECTS

Blockchain 3.0, despite its innovations and promise, still faces several challenges to achieve widespread adoption. Technological complexity and the need for common standards for interoperability and governance pose significant obstacles.

Moreover, regulation remains a crucial area to address, as jurisdictions around the world seek to balance innovation with consumer protection and financial security.

Blockchain 3.0 represents an ambitious evolution of blockchain technology, seeking to overcome the limitations of previous versions and extend its impact across diverse sectors. By solving the challenges of scalability, interoperability, sustainability and governance, Blockchain 3.0 is well positioned to play a central role in the digital transformation of society and the global economy.

CONCLUSION

The second chapter of this study explored in depth the innovations and advances that have marked the evolution of blockchain technology, focusing particularly on smart contracts, Ethereum and Hyperledger Fabric. Each of these components has contributed significantly to the expansion of applications and the maturation of blockchain, paving the way for a multitude of new possibilities. Smart contracts have revolutionized the way transactions and agreements are executed, offering reliable automation and enhanced security. Their ability to self-execute and decentralize has eliminated the need for trusted third parties, reducing costs and the risk of human error.

Smart contracts have become a cornerstone of Blockchain 2.0, enabling applications in diverse fields such as decentralized finance, supply chain management and healthcare services. Ethereum, with its Turing-complete programming language and virtual machine (EVM), has enabled developers to create sophisticated decentralized applications. The transition to Ethereum 2.0 promises to solve scalability and energy efficiency issues, further strengthening its leadership position in the blockchain ecosystem. The innovations introduced by Ethereum have spawned numerous projects and stimulated innovation throughout the industry. Hyperledger Fabric, meanwhile, offered a robust solution for businesses looking to exploit blockchain without the constraints of public cryptocurrencies. Its modularity, PBFT consensus mechanism, and ability to manage transaction confidentiality made Hyperledger Fabric the platform of choice for enterprise networks. Use cases in finance, healthcare, and supply chain management demonstrate its potential to transform business operations. Finally, Blockchain 3.0, with its improvements in scalability, interoperability, sustainability and governance, promises to overcome the limitations of previous generations.

New applications in a variety of sectors show that blockchain is much more than just a cryptocurrency technology, and has the potential to revolutionize many aspects of our economy and society. In sum, this chapter has highlighted how innovations in smart contracts, Ethereum and Hyperledger Fabric have laid the foundations for a new era in blockchain technology. These advances continue to push back the boundaries of what's possible, paving the way for a future in which blockchain plays a central role in global digital transformation. The challenges remain numerous, but the prospects offered by Blockchain 3.0 suggest immense potential for even more varied and impactful applications.

CHAPTER 3

"Blockchain's modus operandi."

The Blockchain, although still in its early stages of development, has rapidly become one of the most promising and fertile technologies for the next generation of interactive web-based systems, particularly through the application of smart contracts (Kosba et al., 2016). The essential characteristics of Blockchain are decentralization, transparency and security (Nofer et al., 2017). This technology consists of three main elements: a Blockchain based on Lhorodatage, a distributed storage mechanism based on a peer-to-peer network, and a consensus mechanism involving distributed nodes.

Blockchain has solved two well-known problems in computer science and mathematics: the Byzantine generals problem, where multiple nodes collaborate to reach a consensus without needing to trust each other, and the double spending problem, where a digital object, such as an electronic currency or a song, cannot be copied and used multiple times (Zyskind et al., 2015). These solutions have paved the way for new applications and wider adoption of Blockchain in a variety of fields.

The structure of the Blockchain is based on a series of sequentially connected blocks, each containing numerous validated transactions. This architecture guarantees that each new piece of information added is linked to the previous one by hash values, thus ensuring data integrity and immutability.

Blockchain uses consensus algorithms to validate and add new blocks, such as proof-of-work (PoW) and proof-of-stake (PoS), each with its own advantages and disadvantages. Blockchain technology is distinguished by its ability to operate in a decentralized environment, eliminating the need for trusted third parties and enabling secure peer-to-peer transactions.

Peer-to-peer (P2P) networks are fundamental to the operation of the Blockchain, enabling every node to participate in the validation and propagation of transactions, thus reinforcing the resilience and security of the network.

Figure 3: The history of blockchain

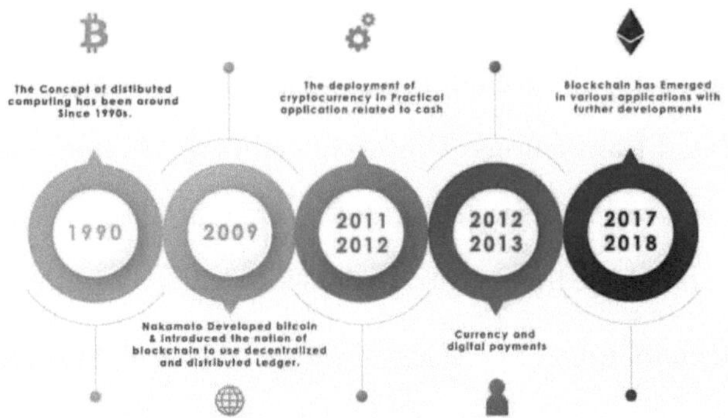

Source : CryptoApe.com

In addition, cryptography plays a crucial role in securing data and transactions on the Blockchain. Asymmetric cryptography, in particular, manages identities and ensures transaction integrity through the use of public and private key pairs. This method ensures that only authorized parties can access and manipulate information stored on the Blockchain.

Consensus algorithms are also essential for maintaining network consistency and security. Proof-of-work, while effective in ensuring security, is criticized for its high energy consumption. Proof-of-stake and other algorithms such as PBFT (Practical Byzantine Fault Tolerance) offer more eco-efficient alternatives that are suited to different types of Blockchain, whether public or private. This chapter explores the inner workings of Blockchain in depth, by examining its fundamental structures, consensus mechanisms, and the cryptographic technologies that ensure its security and efficiency.

It highlights how these elements interact to create a robust, decentralized system capable of transforming not just the financial sector, but a host of other industries too.

Figure 4: The essential characteristics of Blockchain (Source: myself)

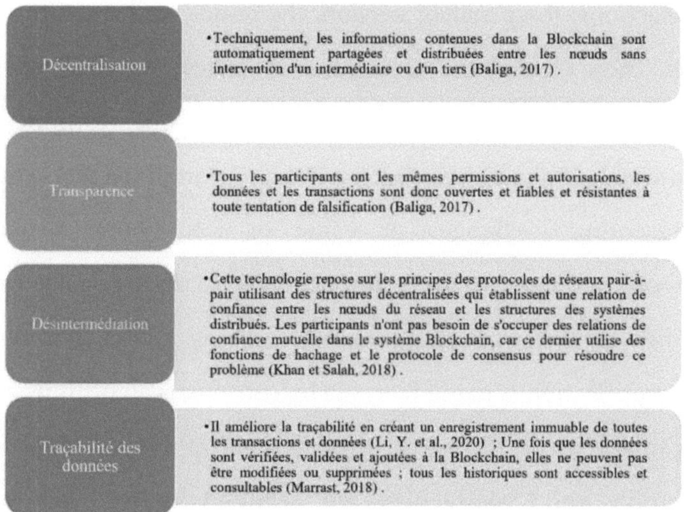

3.1. BLOCKCHAIN STRUCTURE

The structure of the Blockchain is fundamental to understanding how it works and its revolutionary capabilities. At its core, the Blockchain is a series of sequentially connected blocks, each containing a set of validated transactions. This innovative architecture not only ensures data integrity and immutability, but also enhanced transparency and security thanks to its decentralized nature. A block in a blockchain contains several essential components, including a block header and a list of transactions.

The block header includes elements such as the hash of the previous block, the timestamp, the Merkle root of the transactions, the protocol version, the nonce, and the difficulty level. These elements work together to ensure that each block is cryptographically linked to the previous one, forming an unalterable chain. Transactions are at the heart of the Blockchain, representing exchanges of value or information between network participants. Each transaction is securely recorded and verified by the network nodes before being added to a block. This verification is made possible by consensus algorithms such as proof-of-work (PoW) or proof-of-stake (PoS), which ensure that only valid transactions are included in the blockchain. Tokens, often used to represent digital assets, also play a crucial role in Blockchain transactions. They can be exchanged, divided and transferred between network participants, facilitating a wide range of applications from cryptocurrencies to non-fungible tokens (NFTs).

The decentralized nature of the Blockchain is ensured by its peer-to-peer (P2P) architecture, where every node on the network participates in the validation and propagation of transactions. This decentralization eliminates single points of failure and improves network resilience and security. P2P networks can be structured, unstructured or hybrid, each offering different advantages in terms of performance and robustness.

Cryptography plays a vital role in securing data on the Blockchain. Asymmetrical cryptography, using public and private key pairs, ensures that transactions are authentic and data is protected from unauthorized access. This method ensures that only authorized parties can access sensitive information, while enabling transactions to be verified quickly and easily.

Blockchain's structure ingeniously combines elements of cryptography, distributed consensus, and decentralization to create a secure, transparent, and resilient system. Understanding these components and their interaction is essential to fully grasp the potential and applications of this transformative technology.

3.1.1. BLOCK

As previously mentioned, the key phenomenon behind the name Blockchain is the series of blocks, which connect to each other sequentially, like a chain where each block contains numerous transactions, which are validated according to (Antonopoulos, 2014)[34] . The figure below describes an example of a Blockchain, where only a few main entities within the block header are described

Figure 5: Example of a blockchain

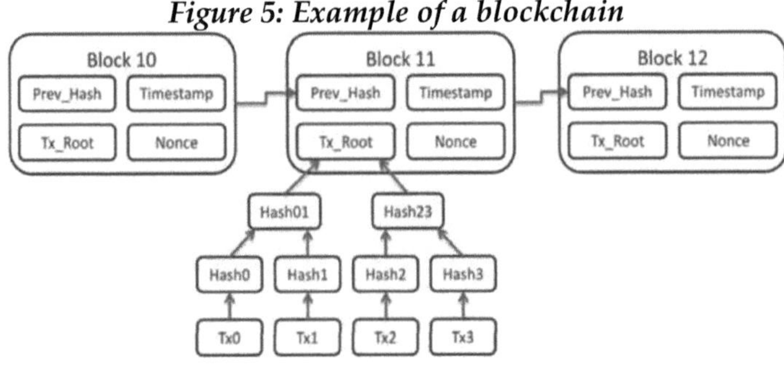

Source : BitConsiel.com

More precisely, *Figure 5* describes the architecture of a block. In addition to the list of transactions contained in the block, the block contains certain fields in the block header:

✓ Prev_Hash: this field can be seen as a parent reference, i.e. a link between a block and the previous block in the chain. All the information contained in the previous block is entered into a hash function to obtain a value, and this value is then assigned to the Prev_Hash field in the new block. In Bitcoin, a 256-bit hash function is used to obtain this value.

- ✓ Timestamp: the time at which the block was found.
- ✓ Tx_Root: this field, also known as the Merkle root, contains the hash value of all validated transactions in the block. As shown in the example in *Figure 5,* all transactions are hashed into one hash value; then they are combined pair by pair and fed into another hash function. This process is repeated until only one entity remains, representing the Merkle root.
- ✓ Version: this field contains the version of the protocol used by the node proposing the block to the chain.
  - ✓ Nonce: this field is used for PoW, which proves the effort a node has made to obtain the right to add its block to the chain. This field will be presented in the next section.
  - ✓ Bits: this field indicates the level of difficulty of the PoW, which will be presented in the next section.

Figure 6: *Block components in a blockchain*

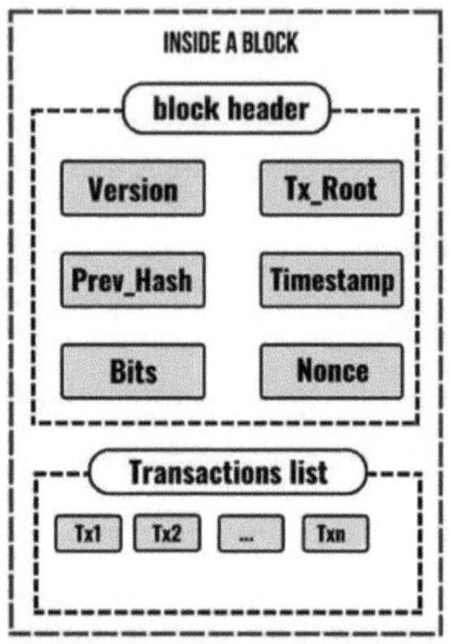

Source : Yhuanlan.com

The block is the basic element for the proper functioning of a blockchain, as data is stored in block form.

3.1.2. Transaction

Data is represented in a blockchain through what is known as a transaction. A transaction consists of n inputs and y outputs. n and y are positive integers.

This means that a transaction cannot be completed in half. A transaction contains the recipient's address. A Token is used as a unit in a transaction. Validating a transaction involves an operating cost, called Gaz.

Figure 7: The components of a transaction

La taille d'un bloc est limitée

Les transactions sont validées par bloc

Le bloc est composé de deux parties, un header et un body

L'arbre de Merkle assure l'integrité

Le header permet le fonctionnement de la Blockchain

Possède le Hash du header précédent

Le body est composé d'une liste de transactions

3.1.3. Token

Tokens, in the context of blockchain, can be defined in terms of several fundamental attributes. Firstly, the transferability of a Token represents its ability to be exchanged between different owners, which is essential for Tokens representing digital assets such as currencies.

Secondly, the subdivisibility of a Token indicates whether it can be broken down into smaller units, enabling greater granularity in transactions. For example, digital currency can be divided into smaller fractions to facilitate low-value transactions.

At the same time, the singleton characteristic attributed to certain

Tokens implies that there is only one instance of that Token, which may be the case for unique objects such as digital works of art. Furthermore, the mint-ability of a Token refers to its ability to issue new units, often carried out by miners in the case of cryptocurrencies.

In addition, the role support feature enables specific roles to be assigned to Tokens in order to regulate access and permissions within a blockchain network, thus contributing to enhanced security and governance. Finally, the burnability of a Token represents the possibility of permanently removing it from the blockchain, which can be applied to exhaustible assets such as fuels or user licenses. These attributes define the different types of Token within the blockchain, offering varied mechanisms for the management, transaction and control of digital assets.

3.1.4. TRANSACTION LIFE CYCLE

Figure 8: The transaction lifecycle

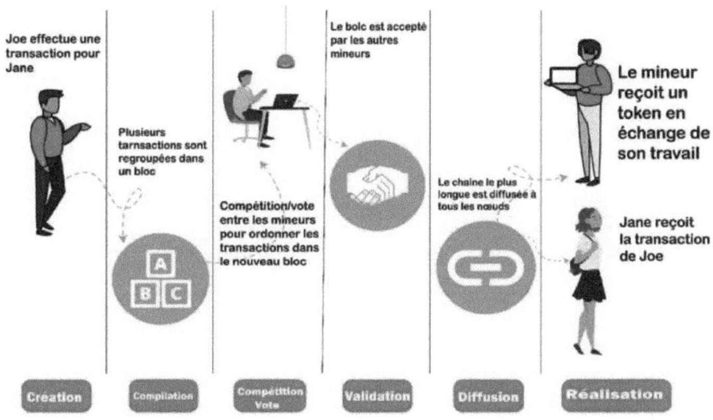

In the life cycle of a transaction on a blockchain, everything starts with its creation. For example, Joe creates a transaction in favor of Jane. Then, several transactions are compiled and grouped together in a block. This block must then go through a consensus process, where network nodes compete or vote to validate the block. Once validated by the community, the block is distributed to all network nodes, ensuring uniform propagation of information. Finally, the transaction is completed when Jane receives the funds and the owner of the validated block is rewarded, in accordance with the specific rules of the blockchain used. This process guarantees the security and reliability of transactions carried out on the blockchain.

3.2. The Peer-to-peer network

First of all, we need to define the different ways in which a network can be organized. We'll see what is meant by the terms "centralized", "decentralized" and "distributed".

Figure 9: Different network architectures

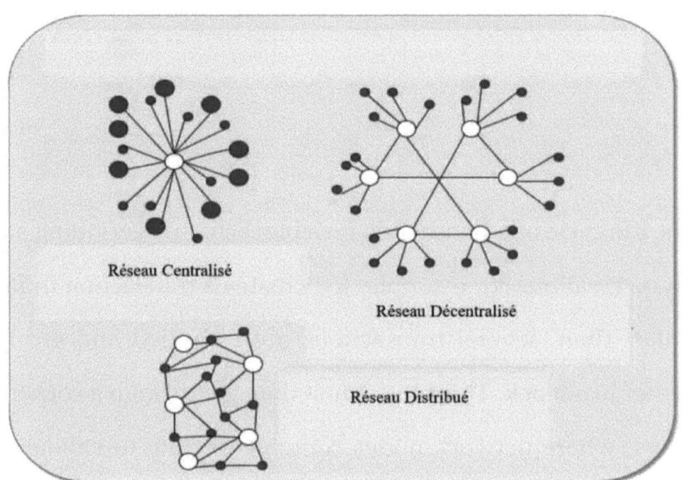

3.2.1. *CENTRALIZED NETWORK*

In a centralized network, the various stations (users) are connected to a single server. From a technical point of view, this is the simplest network to build, but it does present a number of risks, including that of being a single point of failure (SPF).

Indeed, if a single central point fails or malfunctions, the entire service is affected. What's more, if many users try to access this central point, it can quickly become saturated. In addition, as mentioned above, this system is more exposed to computer attacks, as it relies on a single version of the data, making it difficult to detect malicious modifications.

Next, we look at distributed and decentralized networks. The main difference lies in how the decision to add new data is made, and how information is shared between the system's control nodes.

3.2.2. *RESEAU DISTRIBUTES*

Here, a user connects to a server by entering his login details. If the information is correct, the connection is authorized. In this client-server mode, roles are fixed - the client will never be a repository of information, and will always have to acquire it from the server.

Distributed means that processing is shared between several nodes, but decisions can still be made centrally using data from all the nodes in the system. This type of system makes it possible to distribute computing operations that are too heavy for a single network node. However, after partial but complementary processing by each node, a central authority makes a decision.

3.2.3. *DECENTRALIZED NETWORK*

Decentralized systems are also known as distributed computing and distributed databases; independent components that interact with other, different machines that exchange messages to achieve common goals.

As such, the distributed system appears to the end user as an interface or computer. Together, the system can maximize resources and information while avoiding system failures and not affecting service availability.

3.2.4. *PEER-TO-PEER*

Peer-to-peer is a way of connecting computer nodes, and in particular of establishing a decentralized network. Generally speaking, the way we use the Internet operates in a client-server environment (centralized system). In other words, a client makes a request, transmits it to a server which receives and processes it, then sends a response back to the client.

In the late 1990s (Schmandt, C. et al., 1991)[36] , a new type of organization between nodes emerged: peer-to-peer networks. Instead of depending on a central server to which several users connect to make requests, users are directly connected to each other, without intermediaries.

In a peer-to-peer network, each node acts as both client and server, inherently challenging the basis of the client-server environment. Here, each node is able to submit requests to other nodes, while also processing the latter's requests.

To draw an analogy, we can compare the peer-to-peer network to a discussion between scientists, where each participant can answer the questions of his or her colleagues and ask questions of his or her own (Cao, S. et al.,2020).

This type of networking has been popularized by the general public, thanks in particular to file-sharing networks between individuals. This specific form of networking, which does not depend on a single node, is at the heart of blockchain technology. P2P architecture is suitable for a variety of use cases, and can be classified into structured, unstructured and hybrid peer-to-peer networks.

Unstructured peer-to-peer networks are formed by nodes connecting to each other at random, but are less efficient than structured networks. In structured peer-to-peer systems, the nodes are organized, and each node can efficiently search the network for the data it requires.

Hybrid models are in fact a combination of P2P and client-server models, and when compared to structured and unstructured P2P systems, these networks tend to have better overall performance.

Importantly, Blockchain technology exploits peer-to-peer network principles to create a decentralized and secure system, where each node participates in the validation and storage of data. This approach offers numerous advantages in terms of resilience, transparency and information security.

3.3. *CRYPTOGRAPHIC CONCEPTS*

Cryptography is a scientific field whose aim is to ensure the security of messages. This discipline, whose origins date back long before the computer age, has undergone a significant evolution with the advent of modern computing. Cryptography plays an essential role in protecting confidential information and electronic communications.

It relies on advanced mathematical algorithms to encrypt data so that it cannot be read or understood by unauthorized parties. Cryptography also ensures data authenticity and integrity, by making it possible to verify the origin and integrity of messages exchanged.

3.3.1. *SYMMETRICAL CRYPTOGRAPHY*

Symmetrical cryptography involves the creation of a unique encryption and decryption key, which the sender and recipient of the message must share in complete secrecy. The same key is used to encode the message, making it unreadable to all, and to decode it.

Anyone in possession of this key can therefore decrypt a message and read the communication in clear text, as illustrated in *Figure 10.* If A wants to send a simple message to B, they must have a unique encryption key, which they must both keep secret to prevent malicious actors from eavesdropping. If the "I'm ready" message is converted to ciphertext by A using a specific substitution algorithm, B needs to know the substitution change to decrypt the ciphertext when it reaches him. To sum up, the whole process works as follows:

Step 1: Aand B decide on a common key to use.

Step 2: A sends the secret encryption key to B or vice versa.

Step 3: A uses the private key to encrypt the original message.

Step 4: Sent the encrypted message to Jeanne.

Step 5: B uses the secret key to decrypt the message it has already received.

Figure 10: Symmetric encryption

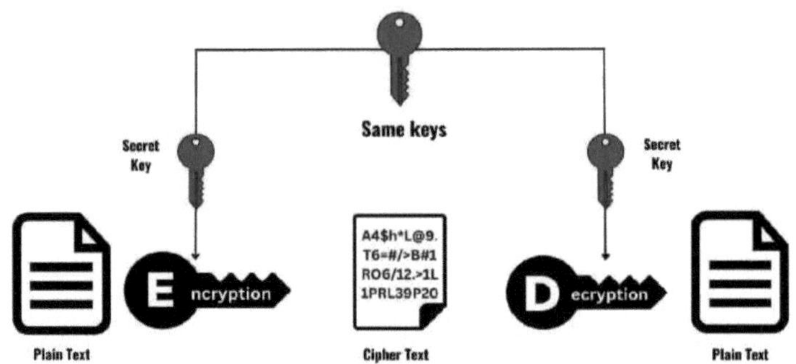

Source : created by the author

By following the above process, A and B communicate privately without fear of anyone lurking in the road. Since only they possess the secret key needed to encrypt and decrypt the message, no third party can intercept the encrypted message. The security of symmetric cryptography relies essentially on the exchange of the secret key between the two communicating parties (N. A. Advani et al., 2019).

This makes it difficult for two distant individuals to agree on a key to use, as it could be intercepted during transmission. A common example of symmetrical cryptography is the "Caesar's code", used by Julius Caesar for his secret correspondence.

This code consists in shifting each letter of the message by a certain number of positions. For example, with an offset of 3, the letter "a" becomes "d", the letter "b" becomes "e", and so on.

This rudimentary encryption was sufficient at the time, given the level of literacy of the population. Caesar's code is a typical example of symmetrical cryptography, also known as secret key cryptography. It is a mono-alphabetic cipher, where an encrypted letter will always give the same letter and vice versa.

There are more advanced symmetrical encryption methods, known as poly-alphabetic ciphers, in which the encryption of a letter can vary according to its position in the message, for example. The famous Enigma machine, used during the Second World War, is an example of symmetrical poly-alphabetic encryption.

Symmetrical cryptography encrypts a message using a key. Once the message has been encrypted, the same key is used to decrypt it and make it readable again.

It should be noted that symmetrical cryptography has both advantages and limitations. Although it is fast and efficient for encryption and decryption, it poses the challenge of secure key distribution between communicating parties.

What's more, if the secret key is compromised, all messages encrypted with that key are also compromised. This is why other forms of cryptography, such as asymmetric cryptography, have been developed to solve these security problems. As an expert consultant, I recommend carefully assessing specific needs and constraints before choosing the cryptographic method best suited to each situation.

3.3.2. *ASYMMETRIC CRYPTOGRAPHY*

This branch of cryptography is much more recent, having emerged in the 1970s with the advent of modern computing. Asymmetric cryptography has fundamental differences from symmetric cryptography, as it relies on the use of two keys rather than one (Maqsood, F., et al., 2017).

Asymmetric cryptography makes it possible not only to encrypt messages, but also to sign them securely. This approach is made possible by the use of one-way and backdoor functions, also known as secret ciphers. Asymmetric cryptography is based on key pairs, comprising a public key and a private key. The public key is accessible to all, and can be used by anyone to encrypt a message intended for its owner. The private key, on the other hand, is kept secret and used exclusively by the recipient to decrypt messages encrypted with the corresponding public key. Asymmetric cryptography also enables messages to be digitally signed. This guarantees authenticity and integrity. The latter is generated using the signatory's private key and can be verified using his or her corresponding public key. One-way functions, also known as hash functions, play an essential role in asymmetric cryptography. They generate a unique, fixed fingerprint for each message, whatever its size. This fingerprint, also known as a hash, can be used to verify data integrity and detect any tampering.

The functions have two main features:

- ✓ Firstly, they are one-way, i.e. it's very easy to calculate the function to go from x to f(x). However, it is extremely complicated to go back to the value x knowing f(x), depending on the length of the key used. Today, the minimum standard for key length used in asymmetric cryptography is 1024 bits. If it takes less than a second to calculate f(x) from x, with a key of this length, returning to x would take around 1500 years on a standard desktop computer.
- ✓ The second property of these functions is that they have a backdoor or secret key. Thus, for a certain specific value of the key used, the time required to go from f(x) to x becomes negligible (of the order of a second).

The key used to encrypt the message is called the public key, while the second key used to exploit the backdoor is called the private key.

***Figure 11**: Asymmetric encryption*

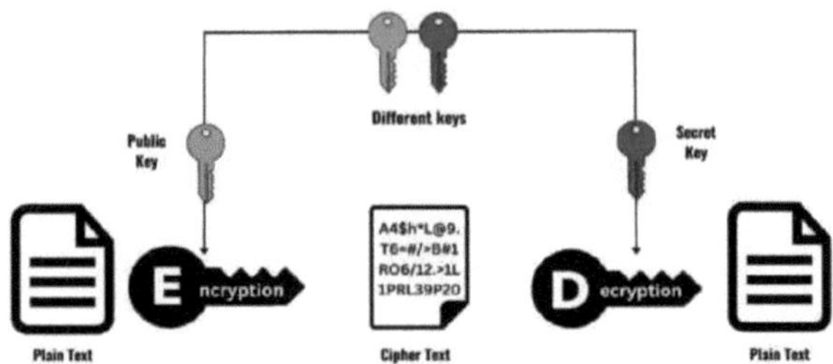

Source : created by the author

Another application of encryption is to use asymmetric cryptography to electronically sign a message. You can use your private key on a message hash. The correspondent can then use the public key to verify that the author of the message is the person who possesses the private key. This also verifies that the message has not been altered between transmission and reception.

***Figure 12**: Message exchange between Alice and Bob intercepted by Eve*

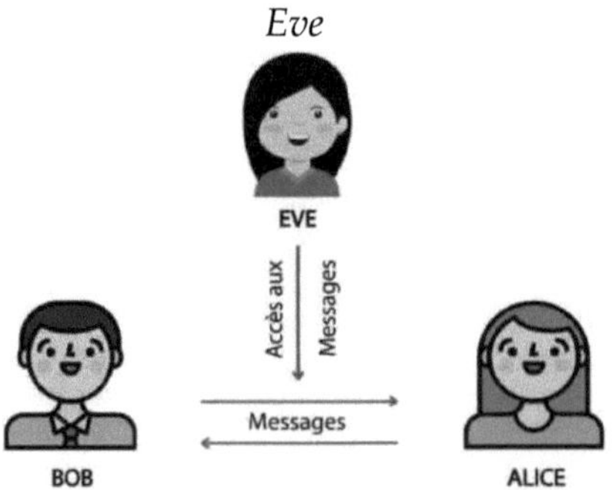

Source : created by the author

3.3.3. *ASYMMETRICAL CRYPTOGRAPHY: ENCRYPTION/DECRYPTION OF MESSAGES*

How can these one-way and backdoor functions be used to encrypt messages? Let's take the example of B (Bob) and A (Alice) who wish to exchange messages securely in the presence of a spy, C (Eve), who would like to know the content of the messages exchanged. Let's suppose that C is able to read all the messages exchanged between B and A.

In asymmetric cryptography, everyone has a private key, shown in red, and a public key, shown in green. As the name suggests, the private key must never be divulged, while the public key can be freely distributed to anyone.

B and A will therefore keep their private keys to themselves, but exchange their public keys, which Eve is able to intercept. A wants to send a "Hello" message to B. He will encrypt it with B's public key, which he received during the key exchange. Encryption with this public key is a one-way function.

In other words, A will be able to encrypt the message with B's public key in a very short time, but it will be impossible (within a reasonable time) to trace the message back to the original, unless he has the associated private key in B's possession.

Figure 13: *Each user has two keys*

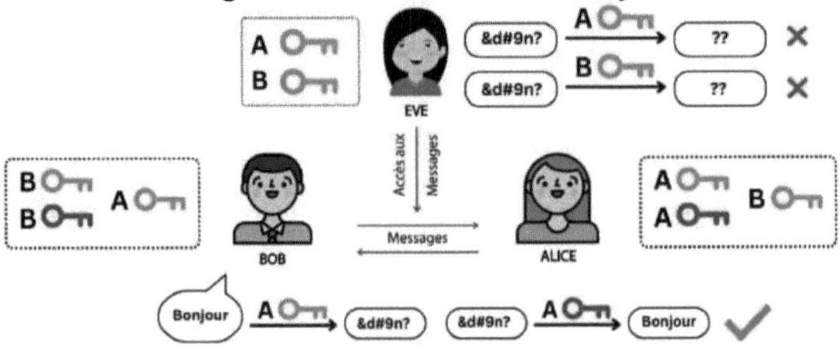

Source : created by the author

Figure 14: *Bob encrypts a message that only Alice can read*

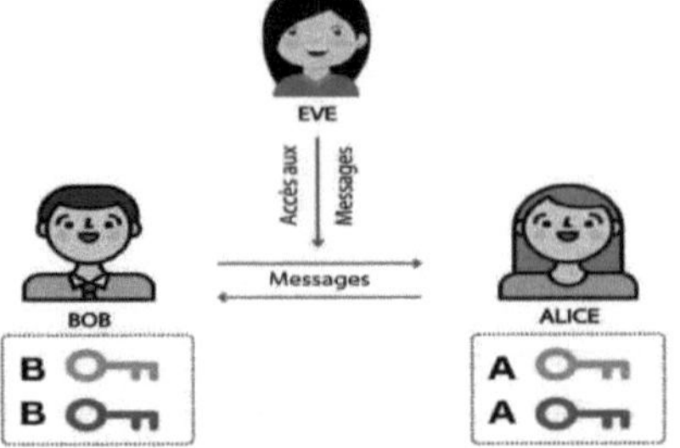

Source : created by the author

On the other hand, B's and A's public keys are of no use to C, who cannot retrieve the original messages. As a result, C intercepts a message that it is unable to decrypt. Only B, with his private key acting as a secret cipher, will be able to decrypt the message very quickly, as shown in Figure 14. Conversely, if B wants to send a message to A, all it has to do is encrypt it using A's public key.

Only A, with his associated private key, will be able to decrypt it. C can intercept the encrypted message, but cannot interpret it. To guarantee the authenticity of the sender, the message is first encrypted using B's public key. The first copy is sent directly to B.

The other copy goes through a hash function and is then encrypted using A's private key before being sent to B. B will first hash the first message received using the same hash function as A to obtain a fingerprint. Next, B runs the second message received through A's public key. If the two match, B is sure that the message comes from A.

3.3.4. *ASYMMETRIC CRYPTOGRAPHY: AUTHOR VERIFICATION AND MESSAGE INTEGRITY*

This technology can also be used to verify the authenticity and origin of a message. To do this, A will, as before, write his message and then encrypt it using B's public key.

It will then run the encrypted message through a hash algorithm to obtain its fingerprint. Finally, he will encrypt this fingerprint, this time with his private key. He then sends both messages to B.

In this example, B wants to check that the message has indeed been sent by A and that it has not been modified since it was sent. To do this, it will calculate the hash of the encrypted message. It will also use A's public key on the second message, which A has encrypted with its private key. If the two results are equal, this means two things:

- ✓ The message has been sent by someone who possesses A's private key (so a is normally the only one to have it).
- ✓ The message has not been altered between transmission and reception (remember that the slightest modification of a file results in a total modification of its fingerprint).

Table 1: Differences between symmetric and asymmetric encryption

Key differences	Symmetrical encryption	Asymmetric encryption
Ciphertext size	The ciphertext is smaller than the original plain text file.	The ciphertext is larger than the original plain text file.
Encryption key	Shared between communicating parts	Two distinct keys: public and private
Use of resources	Symmetric key encryption works on the basis of low resource utilization.	Asymmetric encryption is very resource-intensive.
Key length	128- or 256-bit key size.	RSA key 2048 bits or more.
Security	Less secure due to the use of a single key for encryption.	Much more secure because two keys are involved in encryption and decryption.
Number of keys	Symmetrical encryption uses a single key for both encryption and decryption.	Asymmetric encryption uses two keys for encryption and decryption
Privacy	A single key for encryption and decryption presents risks of key compromise.	Two keys are manufactured separately for encryption and decryption, eliminating the need to share a key.
Speed	Symmetric encryption is a fast technique	Asymmetric encryption is slower in terms of speed.
Algorithms	RC4, AES, DES, 3DES and QUAD.	RSA, Diffie-Hellman, ECC algorithms.

Source : -realised -by the author

3.3.5. *THE DIFFERENCE BETWEEN ASYMMETRIC CRYPTOGRAPHY AND symmetric cryptography*

Symmetrical encryption uses a single key that must be shared between those who are to receive the message, while asymmetrical encryption uses a pair of public and private keys to encrypt and decrypt messages during communication. Symmetrical encryption is an ancient technique, while asymmetrical encryption is relatively recent. Asymmetric encryption was introduced to complement the problem inherent in the need to share the key in the symmetric encryption model, by eliminating the need to share the key using a public-private key pair. Asymmetric encryption takes relatively longer than symmetric encryption. When it comes to encryption, the latest schemes are not necessarily the best. Always use the encryption algorithm best suited to the task in hand.

3.3.6. *INTEREST OF ASYMMETRIC CRYPTOGRAPHY IN A BLOCKCHAIN NETWORK.*

Asymmetric cryptography, also known as public key cryptography, plays a crucial role in the operation of blockchain networks. This cryptographic method, which uses a pair of keys - a public key and a private key - makes it possible to secure transactions, manage identities and guarantee data integrity in a decentralized way. Unlike symmetrical cryptography, where a single key is used to encrypt and decrypt information, asymmetrical cryptography offers enhanced security by separating these functions. One of the main benefits of asymmetric cryptography in a blockchain network lies in its ability to ensure the confidentiality and integrity of transactions. Each network participant has a public key, which can be shared freely, and a private key, which must remain secret. When a user wishes to send information or carry out a transaction, he or she encrypts the message with the recipient's public key. Only the holder of the corresponding private key can decrypt the message, thus ensuring that the data can only be read by the intended recipient. Asymmetrical cryptography also makes it possible to verify the authenticity of transactions and messages. By using his private key to sign a message, a user can prove that he is the author. Other participants can verify this signature using the sender's public key. This cryptographic verification ensures that the message has not been altered in transit, and that the sender is who he or she claims to be. In the context of blockchains, asymmetric cryptography is essential for managing identities and access rights. Each network user has a public address derived from his or her public key, serving as a unique identifier. This address is used to receive transactions and interact with the network. Possession of the associated private key enables the user to prove his or her identity and sign transactions, reinforcing security and confidence in the decentralized network.

Asymmetric cryptography also facilitates the implementation of consensus and governance mechanisms in blockchain networks. For example, Proof of Stake (PoS) and Delegated Proof of Stake (DPoS) algorithms rely on the ability of users to prove possession of their tokens and participate in block validation by cryptographically signing their votes or proposals. In conclusion, asymmetric cryptography is a fundamental pillar of blockchain networks, offering robust security, efficient identity management and reliable transaction verification. Its ability to protect information and guarantee the authenticity of communications is essential for the smooth operation and integrity of decentralized systems.

3.3.6.1. *IDENTITY MANAGEMENT ON THE NETWORK*

We're going to illustrate the property of being able to identify the author and integrity of a message. In the Bitcoin blockchain, asymmetric cryptography is used to manage identities on the network. Each account has a public and a private key. The public key can be thought of as a person's Bitcoin address. A Bitcoin address is the equivalent of an account number.

As with bank accounts, it is possible to have several Bitcoin addresses. All Bitcoin addresses are unencrypted on the network, but do not identify the person who owns them. Using asymmetric cryptography and hash functions, the network can guarantee that A and no one else is the originator of the transaction.

If A wants to send a certain amount of money to B, he has to send a message on the Blockchain network with :

- The address at which to debit the funds
- The address to which the funds are to be credited
- The amount of the transaction.

So the main problem to guard against is identity theft. Since addresses are in the clear on the network, it would be very easy to broadcast a message on the network saying that this address must transfer funds to another. However, when a transaction is issued, it must be accompanied by the transaction's fingerprint, encrypted by the sender's private key. For this reason, only the holder of the private key, who is normally the legitimate owner of the account, can issue a transaction. Suppose A wants to transfer 1 bitcoin to B. He broadcasts on the Blockchain network that his address (considered the public key) is to be debited, that B's is to be credited, and gives the amount. He broadcasts the message and the hash passed by his private key.

To verify that the transaction has come from A, the network simply hashes the message on one side, then decrypts the hash sent by A using its public key on the other. If the two fingerprints are equal, we can be sure of two things:

- ✓ A has issued the order,
- ✓ The data contained in the order corresponds to what A wants (in particular, the amount and the recipient).

There are a number of software programs called "wallets" that make it very easy to create Bitcoin addresses and carry out Bitcoin transactions between these addresses. Interestingly, with the exception of a security breach in 2010, the Bitcoin blockchain has never been successfully hacked. The only successful frauds have been attacks that succeeded in stealing the private keys of certain users in order to transfer the victim's funds to themselves. Major attacks on cryptocurrency exchange platforms have also been successful.

3.3.6.2. ACCESS TO INFORMATION STORED ON A BLOCKCHAIN

Let's not forget that a Blockchain is first and foremost a decentralized information medium. As such, data is replicated between numerous Blockchain users. However, this replication is not incompatible with the storage of private or sensitive data. Indeed, by encrypting data using asymmetric cryptography, there is no antagonism to sharing private data insofar as it cannot be read by other users. Only the person with the private key will be able to decrypt and benefit from the data stored on the Blockchain, while enjoying the many security advantages linked to the register's replication and immutability.

3.3.7. BLOCKCHAIN CONSENSUS ALGORITHMS

While decentralized systems offer many advantages, the way in which the system handles collective decision-making is a question that comes to mind very quickly. Indeed, majority voting in a decentralized system can have many flaws. What happens if a user decides to vote for an invalid transaction? This problem is amplified if this user can create several accounts and thus artificially increase his or her number of votes. Finally, what happens if users agree to validate an invalid transaction? It is essential to propose a consensus system that guarantees sufficient security for the network. It wasn't until the advent of blockchain that mathematicians and computer scientists turned their attention to this problem.

Indeed, the arrival of computing in increasingly critical systems, such as aeronautics or medical devices, has prompted researchers to find systems to make such equipment ever more reliable. One solution is to have redundant calculations, i.e. to make a single decision from several potentially divergent calculations. We find the same problem in blockchains, where different users, called nodes in a network, may have different opinions on a decision to be made. These different opinions may be due to a number of reasons, and we'll see how to deal with these contradictions in the next chapter ("Types of faults in a network"). Despite this divergence, the system must remain coherent. The Bitcoin blockchain is said to be resistant to Byzantine faults thanks to a proof-of-work consensus system. This system ensures that only valid transactions are validated and added to the blocks.

3.3.7.1. PoW (Proof of Work)

Proof-of-Work (PoW) is the first Blockchain algorithm introduced into the Blockchain network. A PoW algorithm works by asking network nodes to solve a mathematical problem in order to create the next block and verify the legitimacy of transactions on the network. The flow of the PoW algorithm is illustrated in *Figure 15.* The mathematical problem is solved using the hash function (Back, A., 2002). Hash is a complex, random mathematical formula used to confirm transactions stored in blocks. All the nodes compete to be the first to find a solution by brute force, which requires a large number of attempts.

Whoever is the first to find the solution may have the right to create a new block, which will be added to the platform once it has been verified. The nodes involved in the calculation are called miners, and the process of solving the problem is called mining. The advantage of the proof-of-work algorithm is its high degree of security and decentralization. Its main drawback, however, is its high energy and resource consumption. Miners need a great deal of processing power to find the solution to the difficult mathematical problem of hashing a billion or more nonces. This wastes precious resources (money, energy, space, hardware). It also takes a lot of time. Miners have to examine a large number of nonce values to find the right solution to the problem that needs to be solved to mine the block. In addition, solving this problem will take some time due to the complexity of solving the hash function. Consequently, this algorithm is not suitable for a large, fast-growing network requiring a large number of blocks.

3.3.7.2. *PoS (Proof-of-stake)*

PoS was mentioned in the first Bitcoin *Figure 16* project, but was not used for robustness and other reasons. The first application of PoS is PPCoin. In PoS King, (S. and Nadal, S., 2012)[41] , the digital currency has the concept of money.

Consequently, the age of a coin is its value multiplied by the period following its creation. The more coins a node has, the more rights it can obtain on the network. As a result, coin holders will also receive a certain reward according to the age of the coin.

Figure 15: *POW algorithm*

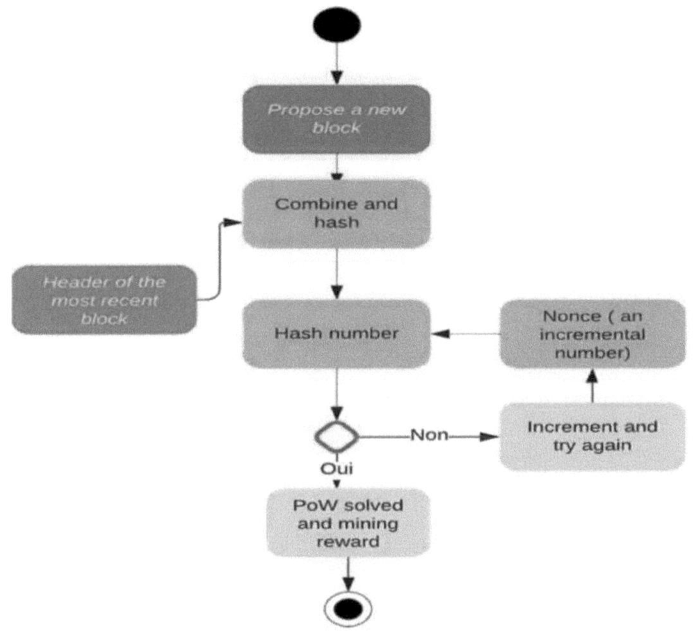

Source : created by the author

In the PPCoin design, mining is also required to obtain accounting rights. The formula is proofhash<the age of the target coin (Mingxiao et al., 2017)[42] . Proofhash is a hash value composed of the weight factor, the unspent output value and the fuzzy sum of the current time.

However, PoS limits the hashing power of each node. Despite the fact that the difficulty of mining is inversely proportional to the age of the coin, PoS encourages coin holders to increase their holding time. Today, with the concept of coin age, the Blockchain is no longer entirely based on proof-of-work. This effectively solves the problem of wasted PoW resources. As a result, Blockchain security using PoS improves as the value of the Blockchain increases. Attackers must then accumulate a large number of coins and hold them long enough to attack the Blockchain.

Figure 16: *Flow chart POS algorithm*

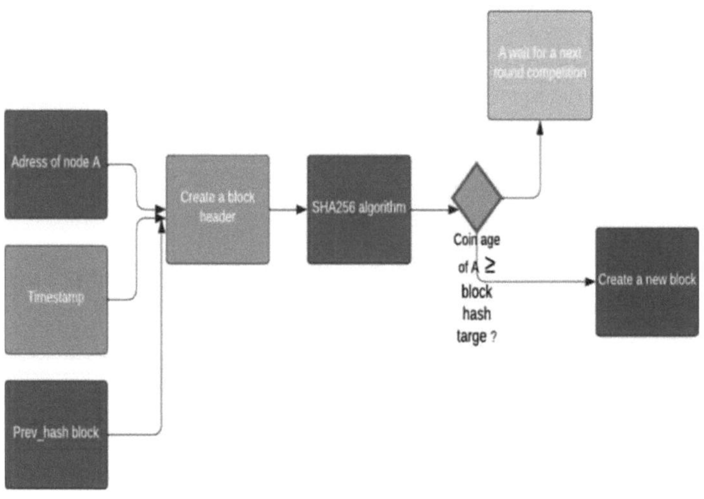

Source : created by the author

This also considerably increases the difficulty of attacks. In addition to PPCoin, there are many other PoS-based currencies, such as Nxt and BlackCion. But they take into account the rights of the nodes and use a random algorithm to assign accounting rights.

3.3.7.3. DPoS (PREUVE D'ENJEUDELEGUEE)

In the Blockchain with DPoS *Figure 17,* each node can select witnesses according to its stake. Over the entire network, the first N witnesses who participate in the campaign and obtain the most votes are entitled to count. The number N of witnesses is defined in such a way that at least 50% of voting stakeholders consider decentralization to be sufficient. In addition, elected witnesses create new blocks one by one as indicated and obtain rewards; on the other hand, witnesses must guarantee an adequate connection time.

Figure 17*: DPOS algorithm*

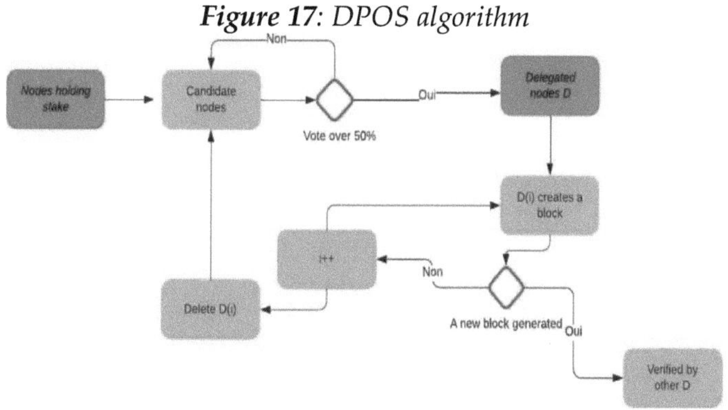

Source : created by the author

If a witness is unable to create the block assigned to it, that block's activity will be moved to the next block and stakeholders will vote for a new witness to replace it. Blockchain using DPoS is more efficient and energy-efficient than PoW and PoS

(Mingxiao et al., 2017)[43] .

3.3.7.4. *PBFT (PRACTICAL BYZANTINE FAULT TOLERANCE)*

In distributed systems, Byzantine fault tolerance can be a good method of resolving transmission errors. However, the first Byzantine system requires exponential operations. Until 1999, the PBFT (Practical Byzantine Fault Tolerance) system was proposed (Castro, M. & Liskov, B.,1999)[44] , and the complexity of the algorithm was reduced to a polynomial level, which significantly improved efficiency (Baliga, 2017)[45] .

The PBFT is composed of five states:

Figure 18: *PBFT states*

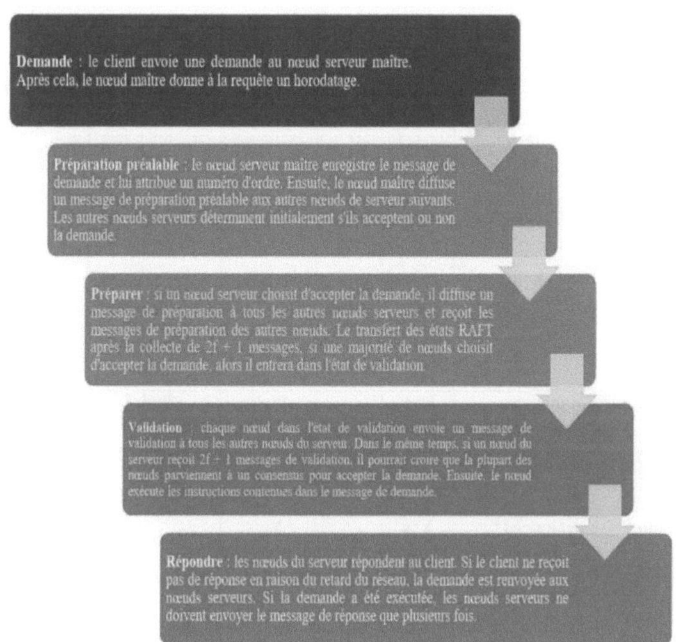

Table 2 below compares the different consensus algorithms discussed in this section.

Table 2: Consensus comparison

	PoW		**PoS**	**DPoS**	**PBFT**
	BTC	ETH	-	-	-
Specific equipment	ASIC	GPU			
Type of blockchain	Public (permissionless)		Hybrid	Public (permissionless)	Private (permissioned)
Transaction purpose	Prob		Prob	Prob	Immediate
Transaction debit	Low		High	High	High
Coin/Token required?	Yes		Yes	Yes	No
Cost of participation	Yes		Yes	Yes	No
Energy consumption?	Yes		No	No	No
Network scalability	High		High	High	Low
Trust model	Without confidence		Without confidence	Without confidence	Semi-confidence
Tolerance of Byzantine faults	<=25%		Depends on Ualgorithme Specific used	Depends on Ualgorithme Specific used	<=33%
Safety level	Very high		Low	High	Medium
Level of decentralization	Medium		High	Very high	Low

Source : created by the author

Despite the fact that numerous consensus algorithms have been introduced and used in the Blockchain, existing consensus algorithms are mainly concerned with the public Blockchain, while the consortium Blockchain receives the least attention.

The consortium blockchain is an authorized blockchain, in which the primary nodes are specified in advance. This means that anyone wishing to access the ledger must be a member of an organization. The consortium blockchain is also made up of known and trusted users.

In general, the PBFT algorithm is widely used in a consortium blockchain, as it considerably improves the blockchain's consensus performance. The number of messages exchanged and processed in the network increases considerably with the number of nodes. Therefore, to overcome the shortcomings of the PBFT algorithm, the clustering method will be added to the PBFT algorithm phase to effectively reduce the communication overhead in the PBFT consensus algorithm.

In the current PBFT algorithm, all nodes must communicate with each other to reach consensus. This entails extensive communication with a large number of nodes, resulting in a higher communication overhead. Therefore, by implementing the grouping method by grouping nodes sequentially, nodes only need to communicate with members of their group instead of all nodes in the network, thus reducing communication overheads.

CONCLUSION

This third chapter explored the inner workings of blockchain in depth, highlighting the fundamental mechanisms that ensure its efficiency and security. Through a detailed analysis of its structure, consensus mechanisms and cryptographic technologies, we have been able to understand how the blockchain succeeds in providing a decentralized, transparent and secure platform for transactions and data. We began by examining the structure of blocks, which form the backbone of the blockchain. Each block contains a set of validated transactions, a cryptographic reference to the previous block, and various other essential information such as the timestamp, Merkle root and nonce. This architecture guarantees the immutability and integrity of the data, making retroactive modification virtually impossible without altering the entire chain. Transactions, the heart of the blockchain, are securely validated and recorded using consensus algorithms such as proof-of-work (PoW) and proof-of-stake (PoS). These mechanisms ensure that all transactions added to the blockchain are legitimate, and that the network operates in a decentralized manner without the intervention of a central authority. Each of these algorithms has its own advantages and challenges, but they all aim to maintain the coherence and security of the network.

Decentralization is a key element that distinguishes blockchain from traditional systems. Thanks to peer-to-peer (P2P) architecture, every node in the network actively participates in the validation and propagation of transactions, eliminating single points of failure and improving overall resilience and security. P2P networks can be structured, unstructured or hybrid, each offering different levels of performance and robustness. Cryptography, particularly asymmetric cryptography, plays a crucial role in securing data and transactions on the blockchain. Using public and private key pairs, asymmetric cryptography ensures the confidentiality of information, enables identity verification and guarantees the integrity of transactions. This level of security is essential to protect users against fraud and attacks. In conclusion, blockchain ingeniously combines the principles of cryptography, distributed consensus and decentralization to create a robust and secure system. Still in its infancy, this technology has already demonstrated its potential to transform not only the financial sector, but a host of other industries too. As it continues to evolve and overcome its current challenges, blockchain is well positioned to play a central role in the digital transformation of our global society and economy.

CHAPTER 4

"Blockchain and distributed ledger technology."

Although the Blockchain and the distributed ledger are similar, there are differences between the two. Blockchain can be classified as a type of distributed ledger, but not every distributed ledger can be classified as a Blockchain, "A Blockchain is a type of distributed ledger, but not all distributed ledgers are Blockchains." (IBM, 2020). However, most companies still use a centralized database with a fixed location. Unlike a centralized database, a distributed ledger is decentralized, eliminating the need for a central authority or intermediary to process, validate or authenticate transactions. A distributed ledger is a database that can reside in multiple locations or with multiple participants, Furthermore in a Blockchain, data is replicated across all nodes in the network, whereas in a distributed ledger, data is only replicated across the nodes that need to know about it." (Coindesk, 2018). Furthermore, these records will only be stored in the ledger once the parties involved have reached a consensus and decentralized peer-to-peer networks On the other hand, Blockchain is a form of distributed ledger that relies on a specific technology. The latter creates an immutable register of records maintained by a decentralized network after a consensus has approved all records. "Distributed ledgers are a type of database distributed across multiple sites, countries or institutions. Blockchain is a type of distributed ledger, where data is stored in blocks that are chained together." (World Economic Forum, 2016). The significant difference between Blockchain and Distributed Ledger Technology (DLT) is the cryptographic signature and linking of groups of records in the ledger that forms a chain. In addition, the public and users have the ability to determine the structure and operation of a Blockchain based on the specific application of the Blockchain, "A distributed ledger is simply a database distributed over a network of computers. A Blockchain is a specific type of distributed ledger that uses cryptographic techniques to secure data and make it tamper-proof." (Investopedia, 2020) We've listed some of the unique aspects of Blockchain and distributed ledgers to better understand the difference between DLT technology and Blockchain.

4.1. *STRUCTURE*

The primary difference between Blockchain technology and distributed ledgers lies in their fundamental structure. A blockchain is generally made up of blocks of data, each block containing a set of validated transactions, a reference to the previous block in the form of a hash, and other important metadata such as the timestamp and nonce. This sequence of blocks creates an unalterable chain, hence the name "blockchain". Each new block is added linearly and sequentially, guaranteeing the integrity and immutability of the data recorded. In contrast, a distributed ledger does not need to adopt a blockchain structure. Distributed ledgers can organize data in different ways according to the needs of the application. A distributed ledger is simply a database distributed across multiple nodes, and this data can be represented in many ways in each ledger (Coindesk, 2018). Some distributed ledgers may use a block structure similar to that of the Blockchain, while others may adopt different formats or not use blocks at all.

4.1.1. STRUCTURE TYPES IN DLTs

1. **Blockchain** :
   - **Linear sequence of blocks**: each block is connected sequentially to the previous block via a hash, forming a continuous chain.
   - **Cryptographic reference**: The reference (hash) to the previous block guarantees that each new transaction is

linked to and dependent on the previous ones, thus ensuring data immutability.

- **Example**: Bitcoin, Ethereum.

2. **Distributed databases** :

- **Data distribution** : Data is distributed over several nodes, but does not follow a block structure. Each node may contain a complete or partial copy of the data.
- **Various consensus mechanisms**: DLTs using this structure can implement various consensus mechanisms adapted to their specific needs.
- **Example**: Corda, Hyperledger Fabric.

4.1.2. Advantages and Disadvantages of Structures

4.1.2.1. Blockchain:

Advantages :

- Immutability: The linear sequence of blocks ensures that once data has been added, it cannot be modified.
- Security: The cryptographic reference and the difficulty of retroactive modification increase security against attacks.
- Transparency: Transactions are visible and traceable in a transparent way.

Disadvantages :

- Limited scalability: linear sequencing can become a bottleneck as transactions increase.

- Energy consumption: Consensus mechanisms such as proof-of-work are energy-intensive.

4.1.2.2. DISTRIBUTED DATABASES :

Advantages :

- Flexibility: Data can be organized and accessed in a variety of ways, depending on the needs of the application.
- Adaptability: Enables the implementation of varied and specific consensus mechanisms.

Disadvantages :

- Consistency: Ensuring data consistency between nodes can be a challenge, especially in highly distributed networks.
- Transparency: The structure may not offer the same transparency as a traditional blockchain.

4.1.3. APPLICATION PROSPECTS

The various DLT and blockchain structures are tailored to different use cases, depending on specific requirements in terms of security, scalability and transparency. Blockchains, with their sequential and secure structure, are ideal for applications requiring strict immutability, such as financial registers and voting systems. Distributed databases offer a flexibility that can be useful in enterprise environments where different levels of permission and data confidentiality are required, such as in supply chain networks or banking consortia. In conclusion, although blockchain is a specific form of distributed ledger, DLTs offer a diversity of structures that enable this technology to be adapted to a wide range of applications. Each structure has its own advantages and disadvantages, and the choice of structure depends on the specific requirements of the project and the environment in which it will be deployed.

4.2. *SEQUENCE*

All blocks in Blockchain technology are in a particular sequence. However, a distributed ledger doesn't need a specific sequence of data. As stated in the article by Tapscott and Tapscott (2016), "all blocks in a Blockchain are in a specific sequence, linked to each other in a chain" (Tapscott & Tapscott, 2016). In contrast, "a distributed ledger does not require a specific sequence of data" (Li et al., 2017). Instead, different nodes in the network can maintain different views of the ledger, with changes and updates propagated across the network in real time. The particular sequence of blocks in a blockchain is essential to ensure data integrity and transparency. Each block contains a cryptographic hash of the previous block, creating an unalterable chain where any attempt to modify a previous block would invalidate all subsequent blocks. This linear sequence ensures that every transaction is recorded permanently and chronologically, offering complete traceability and data verifiability.

4.2.1. THE IMPORTANCE OF SEQUENCE IN BLOCKCHAIN

The sequence of blocks in a blockchain plays several crucial roles:

1. **Immutability and security**: The block sequence ensures that data cannot be altered once it has been added to the blockchain. Any attempt at modification would be immediately detected, as it would break the cryptographic chain.
2. **Traceability**: Transactions are recorded in chronological order, enabling all transactions to be traced back to their origin. This is particularly important for applications requiring complete and transparent auditing, such as financial records or supply chains.
3. **Consensus and Validation**: The sequence of blocks is also

crucial to consensus mechanisms. Miners or validators must work on the last block in the chain, and any attempt to fork or modify it earlier would be immediately rejected by the network.

4.2.2. *FLEXIBILITY OF DISTRIBUTED LEDGERS*

In contrast, distributed ledgers offer greater flexibility in terms of data structure and sequence. Since they do not require a specific sequence, DLTs can be designed for applications where speed and scalability are more important than strict immutability.

1. **Real-time updates**: In a distributed ledger, nodes can maintain different views of the ledger and propagate updates in real time. This allows greater flexibility in data processing and can improve network performance and scalability.
2. **Resilience**: DLT systems can be more resilient to local failures and attacks, as they are not dependent on a single sequence of data. Nodes can continue to operate and update the ledger independently of other nodes.
3. **Diverse applications**: This flexibility makes DLTs particularly well-suited to diverse applications such as

identity management, real-time payment networks, and asset management systems where strict sequencing is not required.

4.2.3. *Use cases and benefits*

The strict sequence of blocks in a blockchain is particularly advantageous for applications requiring high security and immutable traceability. In land registries, for example, the blockchain can ensure that every property transaction is permanently and verifiably recorded. In voting systems, it can ensure that every vote is counted and recorded without the possibility of fraud. On the other hand, distributed ledgers can offer efficient solutions for systems requiring rapid and flexible data updating. For example, in interbank payment networks, the ability to update account balances in real time without following a strict sequence can improve efficiency and reduce processing times. In conclusion, while block sequencing is a distinctive and advantageous feature of blockchain, distributed ledgers offer a structural flexibility that may be more suited to certain applications. The choice between a blockchain and a DLT depends on specific requirements in terms of security, traceability, scalability and system performance.

4.3. *PROOF OF WORK (PREUVE DE TRAVAIL)*

In most cases, blockchains use the Proof of Work (PoW) mechanism. There are other mechanisms, however, but they are generally power-hungry. The distributed ledger, on the other hand, does not require this type of consensus, which makes it more scalable. Proof-of-work adds a significant difference between the distributed ledger and the Blockchain, it's a consensus mechanism that involves solving a complex mathematical puzzle to add a new block to the Blockchain, and is designed to be difficult and resource-intensive to discourage malicious actors from attempting to manipulate the network. According to the original Bitcoin white paper, "proof-of-work consists of looking for a value that, when hashed, as with SHA-256, starts with a certain number of zero bits" (Nakamoto, 2008); the higher the number of zero bits, the harder the puzzle. To guarantee the addition of new blocks to the Blockchain, this requires "substantial computational effort" (Bonneau et al., 2015). This computational effort is provided by the "miners" who compete to be the first to solve the riddle and add a new block to the Blockchain.

4.3.1. *How Proof of Work works*

The proof-of-work process works as follows:

1. **Block proposal**: Miners collect unconfirmed transactions and organize them into a candidate block.
2. **Solving the puzzle**: Miners need to find a nonce (an arbitrary number) which, when combined with the block data and passed through a hash function (such as SHA-256), produces a hash that starts with a certain number of zero bits.
3. **Diffusion and Validation**: Once a miner has found the correct nonce, he diffuses the validated block to the entire network. The other nodes quickly check that the block is valid.
4. **Addition to the Ledger:** If the block is validated by the majority of nodes, it is added to the blockchain and the miner is rewarded with cryptocurrencies.

4.3.2. *Advantages and disadvantages of Proof of Work*

Advantages :

- **Security**: The high cost and difficulty of the puzzle make manipulation of the blockchain costly and impractical for attackers.

- **Decentralization**: PoW enables broad participation in the validation process, distributing validation power across numerous miners.
- **Resilience**: decentralized structure and data redundancy ensure resilience in the face of breakdowns or attacks.

Disadvantages :

- **Energy consumption**: PoW is extremely energy-intensive, requiring a great deal of computing power to solve the puzzles.
- **Scalability**: The difficulty of the puzzles and the time needed to solve them can limit the number of transactions processed per second.
- **Centralized mining**: The high cost of mining equipment can lead to centralization, where only players with significant resources can afford to participate effectively.

4.3.3. *Alternatives to Proof of Work*

To overcome the limitations of proof of work, several other consensus mechanisms have been developed:

1. **Proof of Stake (PoS)**:

- Instead of solving puzzles, validators are chosen to create new blocks according to the number of tokens they hold and are willing to "stake". This mechanism consumes much less energy than PoW.

- **Examples**: Ethereum 2.0, Cardano.

2. **Delegated Proof of Stake (DPoS)**:

- Token holders elect a small number of delegates to validate transactions and create new blocks. This model improves efficiency and reduces energy consumption, while maintaining a certain level of decentralization.

. **Examples**: EOS, TRON.

3. **Proof of Authority (PoA)**:

- A small number of reliable, verified validators are chosen to create new blocks. This model is often used in private blockchains or consortia where complete decentralization is not required.

- **Examples**: VeChain, POA Network.

4. **Proof of Capacity (PoC)** :

- Miners allocate disk space for hash data storage, and blocks are created according to the space available. This model reduces energy consumption compared with PoW.

- **Examples**: Burstcoin, Chia.

4.3.4. *Perspectives of Use and Future Developments*

As concerns about energy consumption and proof-of-work scalability continue to grow, many blockchain projects are exploring and adopting these alternative consensus mechanisms. Ethereum's transition to Ethereum 2.0, using PoS, is one of the most prominent examples of this evolution. Blockchain companies and developers are constantly seeking to balance security, decentralization and energy efficiency to create more sustainable and scalable systems. In conclusion, although proof-of-work played a crucial role in securing early blockchains like Bitcoin, the challenges it poses in terms of energy consumption and scalability are driving the community to explore and adopt more innovative and efficient consensus mechanisms. Distributed ledgers, by not being tied to proof-of-work, offer greater flexibility and scalability, enabling wider and more diverse adoption of blockchain technology across diverse sectors.

4.4. *Implementations and Implementation*

Implementation is a key point to consider when understanding the differences between Blockchain and distributed ledger. Blockchain has many real-life implementations, as it grows in popularity and many uses are developed over time. Since many companies are embracing the nature of Blockchain and slowly integrating it into their systems, you'll also find big giants like Amazon, IBM, etc., offering a good Blockchain solution as a service. In comparison, developers have recently begun to dive deep into the core of distributed ledger technology.

Although there are several types of DLT in the world of technology, there are few implementations in practice. However, they are still under development, and we'll start seeing real implementations very soon.

4.4.1. EXAMPLES OF BLOCKCHAIN IMPLEMENTATIONS

1. **Cryptocurrencies** :

- **Bitcoin:** The first and most famous implementation of blockchain technology. Bitcoin proved that the blockchain can secure financial transactions in a decentralized way without a trusted third party.
- **Ethereum**: Beyond simple financial transactions, Ethereum has introduced smart contracts, enabling unprecedented automation and programmability.

2. **Financial Services** :

- **Ripple**: Used for fast, secure cross-border payments, offering an efficient alternative to traditional transfer systems.

. **JPM Coin**: A cryptocurrency developed by JPMorgan Chase to facilitate interbank transactions and reduce transaction costs.

3. **Supply Chain** :

- **IBM Food Trust**: Using blockchain to ensure food traceability from field to plate, improving transparency and food safety.
- **VeChain**: Enables the traceability of luxury goods, medicines and other critical items to ensure their authenticity and provenance.

4. **Voting systems** :

- **Voatz**: A blockchain-based mobile voting app used to secure and verify votes in elections.
- **FollowMyVote**: Provides a secure and transparent online voting solution thanks to blockchain technology.

5. **Health** :

- **MedRec**: Uses blockchain to manage electronic health records, ensuring the confidentiality and security of patient data.
- **PharmaLedger**: A blockchain consortium for the pharmaceutical industry aimed at improving drug traceability and combating counterfeiting.

4.4.2. *Examples of Distributed Ledger Implementations (DLT)*

1. **FinTech**:

- **Corda**: A DLT platform designed for companies to manage financial transactions securely and efficiently. Used by banking consortia for interbank settlements and financial contracts.
- **Quorum**: A permisioned version of the Ethereum blockchain developed by JPMorgan, used for financial applications requiring a high degree of confidentiality.

2. **Logistics and Supply Chain** :

- **Hyperledger Fabric**: A modular DLT infrastructure supported by the Linux Foundation. Used for logistics applications, enabling transparent and secure supply chain management.
- **TradeLens**: A DLT-based supply chain management platform developed by IBM and Maersk, aimed at improving the transparency and efficiency of global logistics processes.

3. **Energy** :

- **Energy Web Foundation**: Uses DLT to manage renewable energy transactions and green certificates, facilitating peer-to-peer energy trading between producers and consumers.

- **Power Ledger**: A DLT platform for decentralized solar energy trading, increasing the efficiency and transparency of energy transactions.

4. **Identity and authentication** :

- **Sovrin**: Uses DLT to provide self-sovereign digital identities, enabling individuals to control their own identification data.
- **uPort**: A DLT platform for digital identity management that enables users to own and control their personal identities.

4.4.3. *CHALLENGES AND OPPORTUNITIES FOR IMPLEMENTATION*

Challenges :

- **Interoperability**: Ensuring that different blockchains and DLTs can interact seamlessly remains a major challenge.
- **Regulation**: Regulations around the use of blockchain and DLT are still evolving, creating uncertainty for businesses.
- **Scalability**: Some blockchain implementations, particularly those based on proof-of-work, encounter scalability problems that may limit their adoption.

Opportunities :

- **Innovation**: blockchain technology and DLT offer fertile ground for innovation, paving the way for new applications and business models.

- **Efficiency**: By reducing intermediaries and automating processes through smart contracts, blockchain and DLTs can significantly improve operational efficiency.
- **Transparency and security**: the immutability and traceability offered by blockchain can improve transparency and security in a variety of sectors, boosting consumer and business partner confidence.

In conclusion, while blockchain has already proven itself in many applications, distributed ledgers also have considerable potential to transform a variety of sectors. As these technologies continue to develop and mature, we're likely to see increased adoption and deeper integration into existing systems, bringing significant improvements in terms of transparency, security and efficiency.

4.5. *TOKENS*

Distributed ledger technology (DLT) does not require the use of tokens or digital currencies. However, tokens can be useful for specific functions, such as spam blocking or access management. For example, DLT systems can implement tokens to limit malicious actions by imposing costs for certain operations, thus ensuring security and proper use of the network. In Blockchain technology, on the other hand, tokens play a fundamental role. Anyone can theoretically manage a node in a blockchain, but operating an entire node requires considerable network infrastructure and can be difficult to manage without the economic incentives provided by tokens. Tokens are used to represent value, ownership or access rights within the network. They act as incentive mechanisms for participants to validate transactions and secure the network. Tokens can represent a wide range of assets. Cryptocurrencies such as Bitcoin are the best-known examples, but tokens can also represent digital assets such as real estate, works of art or company shares. These digital representations enable fluid and secure transactions, ensuring the traceability and authenticity of the assets exchanged. According to Swan (2015), tokens are considered a "key innovation" in the Blockchain space, opening up new possibilities for digital transactions and asset management.

Table 3: Comparison between blockchain and distributed ledger

	Distributed Leger	Blockchain
Block Structure	It's a distributed database on different nodes, but the data can be different on each node.	This is a distributed database on different nodes with the same data.
Proof of work	It is comparatively more scalable, as it does not require proof of work.	It is a subset of distributed ledgers, but offers additional functionality beyond the scope of traditional DLTs.
Common uses	Used in various fields such as finance, logistics, digital identity, etc.	Mainly used for cryptocurrencies, but also being explored in other sectors
Tokens	No need to have tokens or any other currency on the network	There's a kind of token economy

Source : created by the author

Tokens in a blockchain are not limited to their function as currency. They can also be used for network governance, enabling token holders to vote on important decisions and protocol updates. This decentralized token governance function ensures that decisions are taken collectively and democratically, avoiding the centralization of power.

In conclusion, although distributed ledger technology can function without tokens, modern blockchain relies heavily on the token economy. These tokens not only facilitate transactions; they also play a crucial role in incentivizing participants, securing the network and decentralized governance. As blockchain technology evolves, the use of tokens continues to diversify, moving beyond cryptocurrencies to include a variety of digital assets and innovative new applications.

CONCLUSION

This fourth chapter has clarified the essential distinctions between blockchain and Distributed Ledger Technology (DLT). Although often confused, these two technologies present fundamental differences in terms of structure, data sequence, consensus mechanisms, implementation, and the use of tokens. Blockchain, as a specific form of DLT, is distinguished by the use of sequentially linked blocks of data, secured by cryptographic signatures. This structure enables the creation of an immutable and transparent register, validated by consensus mechanisms such as proof-of-work (PoW) or proof-of-stake (PoS). DLTs, on the other hand, can adopt a variety of structures and do not necessarily require a data sequence or energy-intensive consensus mechanisms, making them more adaptable and often more scalable for certain applications. We have also examined the importance of tokens in blockchain networks, which serve as internal economic mechanisms to encourage participation and maintain network security. DLTs, however, do not necessarily require the use of tokens, which can simplify their implementation and adoption in contexts where tokens are not necessary. In terms of implementation, blockchain has already proven its usefulness in many areas, including cryptocurrencies, supply chain management, and secure voting systems. DLTs, although still under development, show promising potential in a variety of sectors such as finance, logistics, and digital identity, offering decentralized and secure solutions without the specific constraints of blockchain. In conclusion, although blockchain and DLT share common principles of decentralization and security, their structural and functional differences open up diverse perspectives for their use. Blockchain continues to revolutionize digital transactions and trust systems, while DLTs offer greater flexibility and adaptability to meet a wide range of industrial and commercial needs. Together, these technologies are shaping a future in which decentralized systems will play a central role in the digital transformation of our society.

CHAPTER 5

"Blockchain and Big data."

Blockchain and Big Data represent two of the most significant technological advances of recent decades.

Each offers unique benefits and presents particular challenges. This chapter explores how the integration of Blockchain and Big Data can transform various sectors, improve operational efficiency, and solve some of the most pressing problems associated with data management and analysis.

Blockchain, as a decentralized, immutable and secure technology, offers robust solutions for data verification and protection. Big Data, on the other hand, encompasses the vast quantities of data generated daily by individuals and businesses, and requires sophisticated techniques for analysis and value extraction. The intersection of these two technologies promises revolutionary advances in various fields, including finance, healthcare, supply chain and many others.

Big Data has grown exponentially with the advent of information and communication technologies. It refers to the massive digital data generated every day by individuals, organizations and connected objects. According to estimates by the International Data Corporation, data generated worldwide is expected to reach a volume of 175 zettabytes by 2025 (IDC, 2018). This data explosion offers numerous opportunities for businesses and organizations, particularly in terms of decision-making, competitiveness and innovation.

Thanks to the analysis of massive data, it is possible to detect trends, correlations and hidden patterns that often escape traditional analysis methods (Chen & Zhang, 2014). Big Data also makes it possible to tailor services and products to customers' needs and preferences, improving the user experience (Jung et al., 2018).

However, despite its benefits, Big Data also poses significant challenges and limitations, particularly in terms of privacy and personal data management (Kitchin, 2014).

Alongside the emergence of Blockchain, Big Data has matured and established a strong market position. Every connected device, ubiquitous sensors and the Internet of Things (IoT) generate continuous streams of data, exponentially increasing the volume of data available. By 2020, the volume of useful data was expected to exceed 16 zettabytes (Turner et al., 2014). Effective management of this information and knowledge extraction are now seen as key competitive advantages.

This chapter examines how the combination of Blockchain and Big Data can improve data reliability, security and management. It also explores the potential applications of this integration in various sectors, and discusses the challenges and opportunities this presents for businesses and organizations. By integrating these two technologies, it is possible to create innovative and sustainable solutions that leverage the strengths of each to maximize benefits and minimize risks.

5.1. *BIG DATA*

Big Data is a field that has grown exponentially in recent years, with the advent of information and communication technologies. It refers to all the massive digital data generated every day by individuals, organizations and connected objects. According to estimates by the International Data Corporation, data generated worldwide is expected to reach a volume of 175 zettabytes by 2025 (IDC, 2018).

Big Data offers numerous opportunities for companies and organizations, particularly in terms of decision-making, competitiveness and innovation. Indeed, thanks to the analysis of massive data, it is possible to detect trends, correlations and hidden patterns that often escape traditional analysis methods (Chen & Zhang, 2014), Big Data also makes it possible to personalize services and products, based on customer needs and preferences, which helps to improve the user experience (Jung et al., 2018). However, despite its benefits, Big Data also poses significant challenges and limitations.

One of the main challenges is related to the protection of individuals' privacy and personal data, which can be used for malicious or discriminatory purposes (Kitchin, 2014). In addition, Big Data requires advanced technical skills and resources, which can represent an obstacle for small businesses and organizations (Manyika et al., 2011).

In short, Big Data is an evolving field, with both benefits and limitations for businesses and organizations. Its effective and responsible management is crucial to maximizing its benefits while minimizing its risks and negative impacts on society.

5.1. Blockchain's impact on Big Data

In parallel with the emergence of Blockchain, Big Data in turn has matured and established a strong market position. Given that every device is online, where sensors are ubiquitous in our world and generate continuous streams of data, where the sheer volume of data offered and consumed on the Internet is increasing by orders of magnitude, where the Internet of Things is producing a digital footprint of our world.

The volume of data is growing exponentially, and it was expected that by 2020 there would be more than 16 zettabytes (16 trillion GB) of useful data (Turner et al. 2014)[61] .

Big data is where innovative technologies offer new ways of extracting value from the tsunami of new information.

The ability to effectively manage information and extract knowledge is now seen as a key competitive advantage. Many organizations base their core business on their ability to collect and analyze information to extract knowledge and ideas.

However, new challenges have emerged: the risks of data circulation, the trustworthiness of data sources, and the inefficiency of data management and analysis have gradually attracted the attention of researchers. In order to meet these challenges and enhance the security, high availability and credibility of data, numerous studies have been carried out.

In the Big Data sector, three issues are frequently encountered in the use of the data generated, concerning the reliability, security and management of the data that Blockchain can impact on:

1. Reliability: The veracity of data is one of the major challenges when using data whose analysis accuracy is affected. With this Big Data *Figure 19,* quality and accuracy are less verifiable, and the lack of quality and accuracy often results in large volumes and less effective analysis reports. However, the combination of Blockchain and Big Data could be a better solution to this challenge, which always undermines the quality of Big Data analysis results. Blockchain is a distributed, secure and tamper-proof data system based on consensus algorithms. It will also be a repository for data verified and written by the members of this federated ecosystem, while maintaining its consistency and integrity. The ecosystem records all events as the identity holder passes through the various stages, and provides each authorized entity with a verifiable history. In this way, the register can be enriched and verified without limit, and the Blockchain keeps track of every recorded action, eliminating the first exhausting but essential phase of traditional Big Data infrastructure: data cleansing.

2Security: Integrating the Blockchain into the Big Data management structure solves the problem of security, thanks to mathematically verifiable cryptographic signatures. In addition, thanks to the use of the Merkle tree, the Blockchain is immutable, meaning that once data is added to the Blockchain, it can no longer be modified. This property of immutability enables the Blockchain to derive data very reliably. In addition, Blockchain's transparency aligns with the current trend towards open data, in two respects: Legal openness, which means that access to data is legal, as is exploitation and sharing. Then there's technical openness, which means there should be no technical barriers to using the data. In general, open data is only non-personal data (Zyskind et al., 2015). To put it plainly, these data do not include any information about individuals, for obvious privacy reasons. Moreover, the integration of the Blockchain guarantees this rule, since it is based on anonymity and each node of this system is represented by an address.

3. *Management:* Blockchain's ability to store public data securely, transparently and immutably eliminates the possibility of tampering or manipulation by bad actors while promoting data protection and continuous availability since the Blockchain concept does not represent the Single Point Of Failure (SPOF) problem (Lee, 2017).

In the case of secure data distribution and management, the research proposes the IT use of a decentralized Blockchain-based solution for Big Data supporting key functions and discussing its implications for the management and sustainability of digital archives (García- Barriocanal et al., 2017).

Other researchers present a system that can manage customer data securely and in a distributed way using cryptographic primitives as well as a keyword search service (Jiang et al., 2020). Another stresses the importance of trust in Big data and presents a secure, Blockchain-based data sharing system (Yue et al., 2017).

In collaboration with a German bank, a group of researchers is following scientific research methods to design, implement and evaluate Blockchain prototypes used to manage workflow between organizations. The results are encouraging, demonstrating the potential of Blockchain as an infrastructure for managing inter-organizational workflow (Fridgen et al., N. D.).

***Figure 19**: The 8Vs of Big Data (source: myself)*

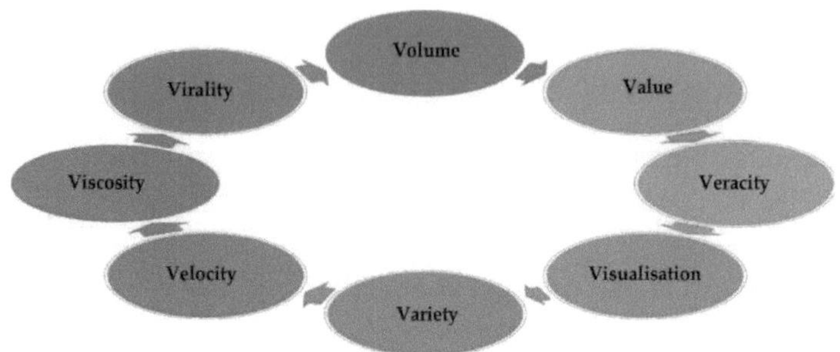

Big Data a stage dependent on Blockchain: an idea that can impact Big Data emphatically to locate an increasingly secure arrangement of information circulating on a network and to guarantee the security, management and veracity of the information to guarantee its value (Kassou, M. et al., 2020). One of the technical aspects of combining these technologies is the use of distributed storage and processing. Big data often requires distributed storage and processing systems, such as Apache Hadoop or Apache Spark, to handle large volumes of data. Similarly, blockchain relies on distributed registers to store and verify transactions across a network of nodes.

***Figure 20**: Classic architecture for Big Data management*

Source : created by the author

When integrating big data and blockchain, it's important to ensure that these distributed systems can work together seamlessly. This requires a careful examination of data formats, storage architectures and processing frameworks several application trials have seen the light such as those applied in the :

Financial sector: the combination of big data and blockchain can enable faster and more secure international payments, as well as fraud detection and prevention. 'IBM has revealed that integrating big data with blockchain can help reduce data reconciliation times by up to 80% (IBM, 2018).

For its part, Accenture has revealed that blockchain can reduce the cost of cross-border payments by up to 40% (Accenture,

2018).

Supply chain management: The combination of big data and blockchain can increase transparency and product traceability, improving efficiency and reducing costs. The University of Arkansas has revealed that the use of blockchain and big data in the food industry can reduce the time needed to trace contaminated food products from days to just seconds (University of Arkansas, 2019).

Energy sector: The combination of big data and blockchain can improve the efficiency and security of energy trading and management. According to the Applied Energy journal publication blockchain can enable more efficient and secure peer-to-peer energy exchanges, which can reduce energy costs and carbon emissions (Zheng et al., 2018).

In turn, the Energy Web Foundation has shown that blockchain can improve renewable energy certificate (REC) markets by increasing transparency and reducing transaction costs (Energy Web Foundation, 2021).

Retail sector: the combination of big data and blockchain can improve supply chain transparency and consumer confidence. According to a revelation in the Journal of Retailing and

Consumer, blockchain can enable more transparent and secure supply chains, which can boost consumer confidence and increase sales (Liu et al., 2020).

Deloitte has shown that blockchain can help retailers reduce supply chain costs by up to 10% (Deloitte, 2019).

Government sector: the combination of big data and blockchain can improve government services and increase transparency. A study published in the Journal of Government Information found that blockchain can enable more secure and efficient government services, such as voting and identity management (Choudhury & Bandyopadhyay, 2019)[70] .

Another study by the World Economic Forum has shown that blockchain can help governments reduce corruption and increase trust in public institutions (World Economic Forum, 2018).

Healthcare sector: the combination of big data and blockchain can improve data management, patient privacy and security, as well as overall healthcare efficiency. For example, a study published in the Journal of Medical Systems found that using blockchain to secure patient data in a hospital can reduce the time spent on administrative tasks by 50% (Shin et al., 2020)[71] .

Another study by Philips Healthcare explored the use of

blockchain and big data analytics to create a secure, decentralized platform for storing and sharing medical images. This platform reduced the time taken to share medical images between healthcare providers from days to just minutes (Philips, 2019)[72] .

Conclusion

The convergence of Blockchain and Big Data is opening up unprecedented prospects for various economic and social sectors. Blockchain, with its ability to provide a secure, immutable and decentralized framework for data management, is proving to be an ideal complement to the challenges posed by Big Data, particularly in terms of reliability, security and management. Integrating Blockchain into Big Data systems not only enhances the veracity of data, but also improves protection against malicious manipulation and guarantees full transparency of transactions. What's more, this technological combination can considerably simplify data management by eliminating single points of failure and ensuring continuous availability. The benefits of this synergy are numerous: in the financial sector, it can speed up and secure cross-border payments while reducing transaction costs; in the supply chain, it improves the traceability and efficiency of logistics processes;

in the energy sector, it facilitates more secure and efficient peer-to-peer exchanges; and in the healthcare sector, it protects the confidentiality of patient data while optimizing information-sharing processes.

However, despite these promising advantages, challenges remain. Effective implementation of these technologies requires interdisciplinary collaboration, adaptation to current regulations and a robust technical infrastructure. What's more, it's crucial to ensure that Blockchain and Big Data systems can interact seamlessly, which requires careful consideration of data formats and processing architectures.

In conclusion, the combination of Blockchain and Big Data represents a major breakthrough that can transform the way data is managed and used. By overcoming technical challenges and fully exploiting the potential of these technologies, companies and organizations can not only increase their efficiency and competitiveness, but also contribute to the creation of a more secure and transparent digital environment. This integration paves the way for a new era of innovation in which the management of massive data is both reliable and secure, opening up unexplored horizons for research and industry.

GENERAL CONCLUSION

Blockchain, often referred to as the fifth major evolution in information and communication technologies (ICT), stands out for its unique ability to transform modern digital infrastructures. Its integration with cutting-edge technologies such as Big Data, the Internet of Things (IoT) and cloud computing, opens up vast and unprecedented prospects for many economic and social sectors.

By offering a secure, immutable and decentralized framework for data management, the Blockchain is proving to be an indispensable tool for meeting contemporary challenges in terms of the reliability, security and transparency of digital transactions.

A study conducted by Price Waterhouse Coopers (2018) showed that 84% of companies surveyed planned to use Blockchain at some point in their business. These companies see Blockchain as a disruptive technology capable of reducing costs, improving process efficiency and increasing the transparency and security of transactions. This growing recognition of Blockchain, both by businesses and researchers, bears witness to its revolutionary potential.

Described as a "trust machine" by The Economist in 2015,

Blockchain is often compared to other emerging technologies such as IoT, cloud computing and Big Data. It is distinguished by its unique and innovative application model, combining distributed data storage, decentralized peer-to-peer transactions, automatic consensus mechanisms, information management systems, programmable smart contracts and dynamic encryption algorithms (Kassou et al., 2020).

According to the Gartner report (2016-2017), Blockchain has been ranked among the emerging technologies most prone to inflated expectations. Nevertheless, it continues to prove its worth by enabling bilateral and multi-party transactions in a distributed and decentralized environment, while offering features such as complete network recording, provenance of information and resistance to forgery.

Furthermore, Blockchain's ability to enforce rules in the event of a breach of trust is essential in a society where a large proportion of interactions are based on trust and the enforcement of rules.

The social and economic implications of Blockchain are vast and potentially polarizing, as this technology could profoundly transform the way we structure value-based transactions and society itself (Al-Saqaf & Seidler, 2017). As Blockchain continues to develop and deploy on a large scale, it promises to become a major driving force in the evolution of digital systems and business practices around the world.

In the healthcare sector, for example, Blockchain can ensure the confidentiality and security of patients' medical records, while enabling the rapid and secure sharing of information between healthcare professionals.

In finance, it facilitates faster, less costly cross-border transactions, while reducing the risk of fraud thanks to its decentralized verification mechanisms.

Supply chain management also benefits from Blockchain, which offers complete traceability of products from origin to final destination, improving transparency, reducing fraud and increasing the efficiency of logistics processes.

In terms of governance and public services, Blockchain could revolutionize identity management, the conduct of elections and the delivery of administrative services. By securing data and making processes more transparent, this technology could boost citizens' trust in public institutions and reduce corruption.

However, the adoption of Blockchain is not without its challenges. Companies and governments must overcome technical, regulatory and cultural hurdles to fully integrate this technology. The complexity of implementing Blockchain-based systems requires specialized skills and a robust infrastructure.

In addition, the need for appropriate standards and regulatory frameworks is essential to ensure the smooth and secure adoption of Blockchain on a large scale. In conclusion, Blockchain represents a major evolution in today's technological landscape. Its ability to offer secure, transparent and efficient solutions for managing data and transactions makes it a key technology for the future.

By exploring and fully exploiting the potential of Blockchain, businesses and organizations can not only improve their efficiency and competitiveness, but also contribute to the creation of a more equitable and reliable digital environment. This integration paves the way for a new era of innovation where the management of massive data is both reliable and secure, opening up unexplored horizons for research and industry.

Blockchain and Big Data, when combined, promise to solve some of the most pressing problems of our time. By offering robust solutions for data management and enabling unprecedented transparency in transactions, these technologies can transform business practices and improve consumer and citizen trust.

As we move forward into this new digital age, it's crucial that we continue to explore, innovate and embrace technologies like

Blockchain to create a more connected, secure and transparent world.

In this way, Blockchain and Big Data are not just technological tools, but catalysts for change that have the potential to redefine the foundations of our digital society.

By adopting these technologies with care and responsibility, we can pave the way for a future where data management and transactions are not only more efficient, but also fairer and more secure. In doing so, we are paving the way for a digital revolution that will benefit us all, creating a world where technology serves humanity, not the other way around.

REFERENCES

1) Al-Saqaf, W., & Seidler, N. (2017). Blockchain technology for social impact: opportunities and challenges ahead. Journal of Cyber Policy, 2(3), 338-354.

2) Antonopoulos, A. M. (2014). Mastering Bitcoin: Unlocking Digital Cryptocurrencies. O'Reilly Media, Inc.

3) Back, "Hashcash - A Denial of Service Counter-Measure," in USENIX Technical Conference, 2002.

4) Baliga, A. (2017). Understanding Blockchain consensus models. Persistent, 4(1), 14.

5) Barriocanal, Elena.and al., (2017). Predicting Patterns in Hospital Admission Data. 10.4018/978-1-5225-2607-0.ch013.

6) Bonneau, J., Miller, A., Clark, J., Narayanan, A., Kroll, J. A., & Felten, E. W. (2015). SoK: Research perspectives and challenges for Bitcoin and cryptocurrencies. In 2015 IEEE Symposium on Security and Privacy (pp. 104121).

7) Cao, S.; Wang, J.; Du, X.; Zhang, X.; Qin, X. CEPS: A Cross-Blockchain based Electronic Health Records Privacy-Preserving Scheme. In Proceedings of the ICC 2020-2020 IEEE International Conference on Communications (ICC), Dublin, Ireland, 27 July 2020; pp. 1-6. [CrossRef]

8) Castro, M. & B. Liskov,(1999), "Practical Byzantine fault tolerance," in Symposium on Operating Systems Design and Implementation, pp. 173--186.

9) Chen, H., & Zhang, H. (2014). Data-intensive applications, challenges, techniques and technologies: A survey on Big Data. Information Sciences, 275, 314-347.

10) Choudhury, T., & Bandyopadhyay, S. (2019). Blockchain in government: A systematic literature review. Journal of Government Information, 163(1), 1-15. https://doi.org/10.1016/j.giq.2019.01.001

11) Coindesk (2018). What Is the Difference Between Blockchain and Distributed
Ledger Technology? Retrieved March 14, 2018, from
https:// www.coindesk.com/what-is-the-difference-between-
Blockchain-and- distributed-ledger-technology

12) Eric Hughes is an American mathematician, computer programmer and cypherpunk. He is considered one of the founders of the cypherpunk movement.

13) Fridgen, Gilbert & Guggenmos, Florian & Regal, Christian & Schmidt, Marco (2017). Big Data beats engineering in residential energy performance assessment - a case study.
14) Gartner. (2020). Top 10 Strategic Technology Trends for 2020.
15) Haber, S., & Stornetta, W. S. (1991). How to time-stamp a digital document (pp. 437-455). Springer Berlin Heidelberg. 19 Back, A. (2002). Hashcash-a denial of service counter-measure.
16) IBM. (2020.). What is a Blockchain? Retrieved March, 14, 2020, from https://www.ibm.com/topics/Blockchain
17) IDC. (2018). The growth in connected devices and Big Data will drive the worldwide data total to 175 zettabytes by 2025. Retrieved from https://www.idc.com/getdoc.jsp?containerId=prUS44498218
18) Investopedia. (n.d.). Blockchain Definition. Retrieved July, 2020, from https://www.investopedia.com/terms/b/Blockchain.asp
19) Jiang, S., Cao, J., Wu, H., & Yang, Y. (2020). Fairness-based packing of industrial IoT data in permissioned blockchains. IEEE Transactions on Industrial Informatics, 17(11), 7639-7649.
20) Jung, H., Song, J., & Kim, K. (2018). Personalization in the Big Data era: A review of the state-of-the-art technologies and challenges. Journal of Business Research, 88, 1-11.
21) Kassou, M. and al, (2020). The Blockchain Based Applications. International Journal of Innovation and Modern Applied Science, 3(3), 1-5, 2020 ISSN: 26658984.
22) King, S. & Nadal, S. ,(2012). "PPCoin: Peer-to-Peer Crypto-Currency with Proof-of-Stake."
23) Kitchin, R. (2014). The data revolution: Big Data, open data, data infrastructures and their consequences. Sage Publications.
24) Lamport. L., R. Shostak and M. Pease, (1982), "The Byzantine Generals Problem," Acm Transactions on Programming Languages & Systems, vol. 4, pp. 382-401.
25) Lee, C. K. M., Cao, Y., & Ng, K. H. (2017). Big data analytics for predictive maintenance strategies. In Supply Chain Management in the Big Data Era (pp. 50-74). IGI Global.
26) Li, C.-T.; Shih, D.-H.; Wang, C.-C.; Chen, C.-L.; Lee, C.-C. A Blockchain Based Data Aggregation and Group Authentication Scheme for Electronic Medical System. IEEE Access 2020, 8, 173904-173917. [CrossRef]
27) Li, X., Jiang, P., Chen, T., Luo, X., & Wen, Q. (2017). A survey on the security of Blockchain systems. Future Generation Computer Systems,

82, 1-13.

28) Liu, X.; Wang, Z.; Jin, C.; Li, F.; Li, G. A Blockchain-Based Medical Data Sharing and Protection Scheme. IEEE Access 2019, 7, 118943-118953. [CrossRef]

29) Cipher machine invented by Arthur Scherbius and Richard Ritter in 1918.

30) Manyika, J. et al. (2011). Big Data: The Next Frontier for Innovation, Competition, and Productivity. McKinsey Global Institute.

31) Maqsood, F., Ahmed, M., Ali, M. M., & Shah, M. A. (2017). Cryptography: a comparative analysis for modern techniques. International Journal of Advanced Computer Science and Applications, 8(6).

32) Mingxiao, D. & al., (2017), "A review on consensus algorithm of Blockchain," IEEE International Conference on Systems, Man, and Cybernetics (SMC), Banff, AB, Canada, 2017, pp. 2567-2572, Doi: 10.1109/SMC.2017.8123011.

33) N. A. Advani and A. M. Gonsai, "Performance Analysis of Symmetric Encryption Algorithms for their Encryption and Decryption Time," 2019 6th International Conference on Computing for Sustainable Global Development (INDIACom), New Delhi, India, 2019, pp. 359-362.

34) Nakamoto, S. (2008). Bitcoin: A peer-to-peer electronic cash system. Decentralized business review, 21260.

35) Nakamoto, S. (2008). Bitcoin: A peer-to-peer electronic cash system.

36) Nofer, M., Gomber, P., Hinz, O. et al. Blockchain. Bus Inf Syst Eng 59, 183-187 (2017). https://doi.org/10.1007/s12599-017-0467-3

37) Philips (2019).blockchain-based radiology data sharing platform: Secureand decentralized. Retrieved from https://www.usa.philips.com/healthcare/sites/healthcare/files/2019-05/philips-blockchain-case-study-radiology-data-sharing.pdf

38) Price Water House, (2018), "blockchain is here. What's your next move?" https://www.pwccn.com/Global-blockchain- survey-2018.

39) Schmandt, C., & Casner, S. Information Sciences Institute-University of Southern California.

40) Shin, S., Choi, S., Kim, J., Kim, J., & Cho, K. (2020). Blockchain-based secure EHR system for healthcare data sharing. Journal of Medical Systems, 44(3), 17. https://doi.org/10.1007/s10916-020-01597-3

41) Swan, M. (2015). Blockchain: blueprint for a new economy. O'Reilly Media, Inc.

42) Szabo, N (1998), Bit Gold, https://nakamotoinstitute.org/bit-gold/

43) Tapscott, D. and Tapscott, A. (2016) Blockchain Revolution: How the Technology behind Bitcoin Is Changing Money, Business, and the World. Penguin, New York. https://www.amazon.com/Blockchain-Revolution- Technology
44) The Economist (2015). The trust machine. Retrieved April 2, 2020, from https://www.economist.com/leaders/2015/10/31/the-trust- machine.
45) Turner, V., Gantz, J. F., Reinsel, D., & Minton, S. (2014). The digital universe of opportunities: rich data and the increasing value of the internet of things. Rep. from IDC EMC.
46) Wang, Y.; Zhang, A.; Zhang, P.; Wang, H. Cloud-Assisted EHR Sharing with Security and Privacy Preservation via Consortium Blockchain. IEEE Access 2019, 7, 136704-136719. [CrossRef]
47) Wei. Dai, Y. Wang, Q. Jin and J. Ma. (2016). "An integrated incentive framework for mobile crowdsourced sensing," in Tsinghua Science and Technology, vol.
21, no. 2, pp. 146-156, April 2016, doi: 10.1109/TST.2016.7442498.
48) World Economic Forum. (n.d.). What is a Blockchain and how does it work? Retrieved May, 2016, from https://www.weforum.org/agenda/2016/06/what- is-Blockchain-and-how-does-it-work/
49) Yue, L., Junqin, H., Shengzhi, Q., & Ruijin, W. (2017, August). Big data model of security sharing based on blockchain. In 2017 3rd International Conference on Big Data Computing and Communications (BIGCOM) (pp. 117-121). IEEE.
50) Zheng, Z., Xie, S., Dai, H. N., Chen, X., & Wang, H. (2018). An overview of Blockchain technology: Architecture, consensus, and future trends. In IEEE International Congress on Big Data (pp. 557-564). IEEE. doi: 10.1109/Big DataCongress.2018.00094
51) Zyskind, G., & Nathan, O. (2015, May). Decentralizing privacy: Using Blockchain to protect personal data. In 2015 IEEE Security and Privacy Workshops (pp. 180-184). IEEE.

Printed by Books on Demand GmbH, Norderstedt / Germany